Report on the Mission To Brazil
to
Research the Cultivation and Preparation of Tea

English Translation of

Rapport de M. Guillemin D.-M, Aide de Botanique à au Muséum d'Histoire Naturelle à M. Le Ministre de l'Agriculture et du Commerce sur sa mission au Brésil, ayant pour objet principal des recherches sur les cultures et la préparation du thé et le transport de cet arbuste en France

By Antoine Guillemin, D.M.P.
1839

Brown Dog Press
Asheville, NC, USA 2022

The original texts used to compose this edition are in the public domain and not subject to copyright protection. However, the translations presented here are new creative works and are therefore protected by copyright. Do not reproduce the English language passages in any form or by any electronic means, including information storage and retrieval systems, without permission in writing from the publisher. Brief passages may be quoted in a review.

Please support our efforts by citing this work appropriately.

Thank you for choosing this book. Watch for additional historical titles on drinks culture and other subjects that strike our fancy.

Copyright © 2022
Brown Dog Press
Asheville, NC 28804 USA

browndogpress.com

ALL RIGHTS RESERVED

Typeset in Adobe Caslon Pro
Printed in the United States
Book & Cover Design by Danijela Mijailovic
Cover illustration by Emile Frederick Nicolle

ISBN 978-1-952432-00-2
Library of Congress Cataloging in Publication Data Pending

This work includes the English translation of:

Antoine Guillemin, *Rapport de M.Guillemin D.-M, Aide de Botanique à au Muséum d'Histoire Naturelle à M. Le Ministre de l'Agriculture et du Commerce sur sa mission au Brésil, ayant pour objet principal des recherches sur les cultures et la préparation du thé et le transport de cet arbuste en France* (Paris: Maulde etRenou, 1839).

https://gallica.bnf.fr/ark:/12148/bpt6k9600367g/f1.image

J. A. T. Rendon, *Pequena Memoria*, in C.A Tauney and L. Riedel, *Agricultor Brazileiro Memoria, segunda edição* (Typographia Imperial e Constituciona lde J Villeneuve e Comp. Rio de Janerio, 1839) 137-149.

https://books.google.com/books?id=riRYAAAcAAJ&pg=PP9#v=onepage&q&f=false

L. S. Sacramento, Memoria Economica Sobre a Plantação, Cultura e Prepara do Chá (1825, reprint: Rio de Janiero, Ty p. Revistas dos Tribunaes, 1908).

https://escoladecha.com.br/blog/freileandromemoria-economica-sobredocha/

and:

Antoine Lasèque, *Notice Sur la Vie et les Travaux d'Antoine Guillemin*, D.M.P. Annales Des Sciences Naturelles, Botanique. Series 2, 17 (1842): 287-296.

https://hdl.handle.net/2027/ucm.5324224115?urlappend=%3Bseq=291

*Report on the Mission
to Brazil
to
Research the Cultivation
and Preparation of Tea*
—
1839

Table of Contents

Publisher's Introduction 1

Introduction — Letter to the Minister of Agriculture
and Trade. 7

Preparatory instructions and documents for the trip
— Crossing — Visits to influential people in
Rio de Janeiro 9

Cultivation, Picking and Preparation of Tea at the
Botanical Garden of Rio de Janeiro 13

Steps to Obtain Tea Plants and Seeds 21

Research on the Origin and the Determination of
Wood Dyes, Medicinal Barks and Drugs —
Acquisition of Wood for Construction,
Woodworking, etc. 23

Sowing of Tea 27

Voyage in the Province of São Paulo — Visits to the
Proprietors — Tea Growers — Cultivation and
Preparation of Tea in the Province — Costs of
Cultivation and Production 29

Departure from Saint-Paul — Voyage to Ubatuba
— Information on the Trade of this Country   39

Return to Rio de Janeiro — Fabrication of Cases
for Tea Plants..................................41

Preparation for the Return to France on the Corvette
Heroine — Tea Plants and Seeds obtained from the
Botanic Garden — Care of the Tea Cases by Captain
Cecille — Effect of the Crossing on the Tea Plants. . 45

Arrived at Brest — Difficulties in Transporting Teas
to Paris — Agricultural Excursion in the Finistère
Department — Care Given on Their Arrival
in Paris......................................49

Material Results of the Mission — Tea Plants and
Seeds and Other Products 53

Information on the Climate and Soil of Brazil —
Tea Cultivation — Sowing — Planting — Harvest
Operations Related to the Preparation of Various
Kinds of Tea — Color and Fragrance of Tea.....55

Results of the Above Observations for Tea Culture
in France 63

Addendum: Additional Reports on Tea Production
in Brazil....................................67

Brief Report on the Planting and Cultivation of Tea
by José Arouche de Toledo Rendon 69

Economic Report on the Planting, Cultivation and
Preparation of Tea By Father Leandro do
Sacramento..................................99

Afterword: Notice on the Life and Works of Antoine
Guillemin, D.M.P. 149

Publisher's Introduction

Tea reached the western world in the 1600s after centuries of use in Asia. From that time until the first half of the nineteenth century, tea was produced exclusively in Japan and China. Japan refused trade with Western nations. China began trading with Western countries in the early 1800s but prohibited foreigners from accessing the interior of the country. Information on tea production was tightly guarded while tea itself found a ready market with European consumers. Western tea consumption increased each year to the point that a significant trade imbalance developed between the tea producing monopoly of China and the tea consuming countries in Europe and North America. There were two responses from Western nations – war and local tea production. The English fought the Opium Wars in 1832 and 1856 to force China to accept increased western trade including imports of British grown opium to offset the cost of tea purchased for their home country. Other western countries benefited from access to the open Chinese markets, but they did not have opium or other goods to trade to offset their increasing purchases of tea and other Chinese goods. These nations sought to reduce their tea imports by establishing tea production in their home countries and colonies.

In 1812 Chinese tea growers along with tea plant seeds from the Portuguese colony of Macau were brought to Brazil at the request of the Emperor or Portugal. Demonstration plantings in the Royal Garden in Rio de Janeiro grew in to the plantation, slave labor based tea industry. This system persisted after Brazil became independent from Portugal in 1822. Shortly after this, tea plants were discovered in the British colonies in India leading to the start of the English controlled tea farms in India and Ceylon (modern Sri Lanka). Political disruption in China from the late nineteenth to the mid twentieth century along with the growth of tea industries in India, Ceylon and other English colonies in Africa displaced China as the major supplier of tea to western countries.

France had no colonies after the end of the Napoleonic wars in 1814. The Second French Colonial Empire began in 1830 with the capture of Algeria. In 1838, shortly before the first Opium War, the French Ministry of Agriculture sponsored an expedition to Brazil to study that country's tea industry. The expedition described here was sent to learn the methods of tea plant propagation and tea leaf processing. The French mission was led by the botanist Dr. Antoine Guillemin. Dr. Guillemin wrote a summary report of the expedition for the Minstry of Agricultre. The report was published in the *La Revue agricole: archives de l'agriculture et de la statistique rurale* (The Agricultural Review: Archives of Rural Agriculture and Statistics) in 1839, and then as a separate pamphlet for wider distribution. The text of the pamphlet was used to prepare this edition. The report is presented here for the first time in English translation. In his report, Dr. Guillemin chronicles his journey to Brazil, the methods used to grow and process tea by the plantation owners and their slaves as well as his efforts to bring viable plant material back to France. Dr. Guillemin's work was highly valued. He was awarded the *Légion d'honneur.* However attempts to establish commercial tea gardens in France using plant material from the expedition were ultimately unsuccessful. Dr. Guillemin died three years after the expedition. No further analysis of the information he collected on tea plants during the expedition was published although his report did encourage other nineteenth century botanists and researchers to explore the possibility of growing tea in France. Translations of those works will be presented in future publications.

Dr. Guillemin indicates in his report that he obtained a copy of the 1839 edition of *Agricultor Brazileiro Memoria*. The work contains information on various agricultural crops and practices in use in Brazil at the time of the French expedition. The work includes a lengthy report on tea growing and processing in use at the time of Guilleman's visit. Guilleman indicates that he planned to prepare a translation of the report but neither a reference to a translation nor a translation itself has been located. A new English translation of the report is included in this edition as an Addendum.

Also included is an earlier report from the director of the national botanical gardens that gives information obtained from a Chinese tea maker who was part of the group that established tea production in Brazil.

Finally a friend's touching summary of Dr. Guilleman's life and work written shortly after his death is included. This gives a full picture of the man who sought to bring tea cultivation to France. A bibliography of Guilleman's printed work given in the article is presented here as well.

Brazilian tea production peaked in 1852 at 65,000 lbs. The industry collapsed when slavery was abolished in 1888. Production was reestablished in the 1920s by Japanese immigrants using plant material from Sri Lanka and India. After that second start, the Brazilian industry has grown and thrived. Annual Brazilian production is over 560,000 tons. Harvesting and processing are largely automated now. Much of the tea produced is sold to the United States for use in blends including iced tea products.

At the time Dr. Guillemin prepared his report, low labor costs allowed the major exporting regions to produce less expensive tea. Higher labor costs made tea production impractical in developed economies like France — that is, until recently. Mechanization has reduced the need for human labor. Greater understanding of tea plant biology and identification of plant varieties that grow in what were previously marginal climates has contributed to the spread of tea production into new areas. At the time this translation was prepared, a small commercial farm on Reunion Island, a department of France located in the Indian Ocean east of Madagascar, was producing tea for sale. A viable, although small, industry is now taking hold on the French mainland. Tea educator Jane Pettigrew recently documented current tea growing projects by the French in her article, *The Tea Growers of France*, presented in the January/February 2022 issue of *Tea Time* magazine. She notes that there are twenty growers in the country, including several in Brittany, where Dr. Guillemin felt the tea plant would thrive. These modern growers support one another through the newly formed *Association Nationale pour la Valorisation des Producteurs de Thé Français,* while the larger European tea growing community has joined together in the *Tea Grown in Europe Association.* Today's European tea growers have realized the hope that underpinned Dr. Guillemin's expedition.

The pre-industrial cultivation and processing methods presented in the reports from Dr. Guillemin and Lt. General Rendon are of mostly of historical interest. Experience, observation and advances in agricultural practices, plant science and biochemistry over the past 190 years have demystified and modernized tea production. Economic concerns still determine the success or failure of a tea growing operation. However, modern tea consumers have a growing interest in tea that is the product of their local soils, weather and labor. Hopefully the success of historical producers like the ones discussed in this edition, coupled with the enthusiasm of modern growers will encourage readers to try their hand at producing tea close to their own home. Enjoy!

Jim Karegeannes
Asheville, North Carolina
April 2022

Dr. Guillemin's Letter to the Minister of Agriculture and Trade

Minister,

In your letter of July 28, 1838, you instructed me to go to Rio de Janeiro in order to procure and bring back tea seeds and rooted plants, in quantities large enough to enable a large-scale attempt to cultivate this plant in various parts of France. You have kindly agreed to assign me Mr. Houlet, deputy head gardener of the hot greenhouses of the Museum of Natural History, to help me in the care of the material results of this mission.

When I arrived in Rio de Janeiro, you instructed me to visit the tea plantations and to carefully study all the cultivation processes, as well as those used for the harvesting and preparation of its leaves. In recommending that I should then take care of the means of obtaining the greatest possible quantity of seeds, and at least fifteen or sixteen hundred rooted plants, you intended that we would return to France around June

1839; to this end you asked the Minister of the Navy to arrange for myself and Mr. Houlet to embark on a government ship, and you asked the Minister of the Navy to the Commander of the Naval Forces in Brazil to give me recommendations for facilitating the shipment to France of the crates containing the teas and their seeds. At the same time, the Plenipotentiary Minister of France in Brazil, on the invitation that he would receive from the Minister for Foreign Affairs, was to give me all the facilities that depended on him in order to carry out the mission that you entrusted to me as successfully as possible.

In the course of the campaign which I have just completed, I had the honor of transmitting to you a few details about the start of my operations in my letters of November 17, 1838, February 27, March 22, and May 16, 1839, reserving the right to submit to you a general report on my mission as soon as it had reached its final conclusion. This is what I am hastening to do today, as the cases of tea have arrived in Paris, and you can forward them to the destination of your choice.

Preparatory instructions and documents for the trip — Crossing — Visits to influential people in Rio de Janeiro

As I was not due to embark until the middle of August, I took advantage of the few days I had left in Paris to gather important information about the cultivation, preparation, and trade of tea in various parts of the world. In this respect, Mr. Gaudichaud, of the Academy of Sciences, who during his circumnavigations has visited most of the places where tea is grown extensively enough to make it an object of trade. Mr. Gaudichaud, I say, has been a great help to me. I also owe a great deal to the teachers at the Museum of Natural History, and particularly to Messrs. de Mirbel and Adolphe Brongniart for the advice they gave me. The reading and communication of all the writings that have been published on tea have been generously provided to me by Mr. Benjamin Delessert, member of the Chamber of Deputies and of the Academy of Sciences, who, moreover, has been kind enough to recommend me to his correspondents in Brazil for all my possible needs. Finally, I carefully collected the documents that Mr. Wallich, Superintendent of the Garden of Calcutta, and Mr. Blume, Chief Medical Officer of the Dutch Army in India, had recently published on tea cultivation in Upper Assam and the island of Java.

Anticipating that I could render commerce and science a great service by determining exactly the origin and nature of the various commodities imported from Brazil, I turned to Mr. Guibourt, a professor at the School of Pharmacy, who gave me a notebook of questions to be answered on wood, construction, woodworking, dyeing, gums, resins, balms, etc., of which we only knew the vernacular, and mostly garbled, names.

A new process, devised by Mr. Ward of London, for the transport by sea of living plants from distant countries, had already been successfully tried by Mr. Wallich. I thought this would be a good opportunity to test it in comparison with other processes proposed by traveling botanists. In Brest, I obtained one of the boxes sent by Dr. Wallich and filled it with twenty-four very beautiful varieties of camellias. Another reason for taking these shrubs to Brazil was that I wanted to make them available to the Brazilians, who would help me to carry out my mission fruitfully. While in Brest, I also received a box of young European fruit trees which were also intended as gifts.

I embarked on August 18, aboard the Dordogne, a cargo corvette commanded by Mr. Filhol-Camas; but the ship was held in Brest harbor until August 27. Our crossing was fifty-three days, during which I made observations on the camellias I had on board. I had the advantage of traveling with Mr. Buchet de Martigny, Consul-General in Buenos Aires, to whom I handed over the European fruit trees, because I felt that they would not succeed in the hot climate of Rio, and that they would be better suited to the regions bordering the Plata river, whose climatic conditions are similar to those of our European regions.

When I arrived in Rio de Janeiro, I was eager to visit the people who could give me information about cultivating tea in Brazil. In addition to the official letters you had given me for Baron Rouen, Plenipotentiary Minister of France, I had received other, no less effective letters from Mr. Auguste Saint-Hilaire, who left such honorable memories in the Brazilian Empire, and whose writings enjoy a deserved reputation. Among the persons of distinction to whom Mr. A. de Saint-Hilaire had recommended me, I am pleased to mention the Lisboa family, one of whose members is now President of the Province of São Paulo, and with whom I wanted to have direct and essential relations for the success of my mission. — Mr. Achille Richard, Professor of Botany at the Faculty of Medicine in Paris, had also written on my behalf to Dr. Ildefonso Gomes, who has done me a great service, as you will see in the rest of this report, and to whom I owe my acquaintance with Dr. Riedel, a learned botanist whom the Russian government has entrusted for twenty years with the task of exploring the vast territory of Brazil. — I would be very ungrateful if I did not mention among my protectors, Mr. Theodore Taunay, consul of our nation, and his two brothers, Mr. Charles and Mr. Felix Taunay, one a major in the service of the Brazil, the other president of the Rio Society of Fine Arts. — Finally, I soon got in touch with Doctors Sigaud and Cuissart, French doctors established in Rio, who, having friendly relations with the country's leading figures, facilitated my access to them, and consequently made a powerful contribution to the success of my undertaking.

It would be exceeding the limits of this report, Minister, if I were to tell you the full range of my activities during the first two months of my stay in Rio. I will therefore prune all that relates to natural history itself, to speak to you only about my investigations in relation to tea and commercial substances.

Cultivation, Picking and Preparation of Tea at the Botanical Garden of Rio de Janeiro

Baron Rouen corroborated the information on the cultivation of this shrub in Rio de Janeiro which I had received in France from Mr. Gaudichaud and Mr. Mittre, surgeon of le *Hercule*, who had accompanied His Royal Highness the Prince of Joinville on his journey to Brazil. He first asked me to visit the botanical garden established near the lake of Freytas, directed by Dr. Bernardo Jose de Serpa-Brandao. I went there on October 28, accompanied by Mr. V. Lisboa and Mr. J. Texeira, friends of Mr. A. de Saint-Hilaire, who introduced me to the Director and recommended me to him. The gift I gave him of some of my camellias increased his good will towards me; but I regretted very much that I had not brought him books on botany or horticulture, which I think would have been even more agreeable to him. Consequently, I proposed to write as soon as possible to my friends in Paris to repair this oversight. The rest of my camellias were given by Mr. Rouen, on my behalf, to Mr. Paul Barbosa, Intendant of the Court, to be cultivated in the imperial garden of St. Christopher.

Mr. B. de Serpa Brandao invited me to visit him often, and promised to show me all the details of cultivating, harvesting, and preparing tea.

On October 30, Count Eugene Ney put me in touch with Mr. Alchorne, director of the English Company for the Navigation of the Rio Doce, who was temporarily in Rio de Janeiro, and who gave me important details about the cultivation of tea in Ouropreto (province of Minas Geraes).

I knew from him that it had flourished in the botanical garden of that city, directed by Mr. Vasconcellos. Mr. Alchorne invited me to go and spend a few days with Mr. March, an Englishman living in the Serra dos Orgaos, two days from Rio, where there are tea plantations. Mr. Th. Taunay also gave me letters of recommendation for Granjean de Montigny, a French architect, whose hobbies are devoted to agriculture, and who engaged in the cultivation of tea in his country house near the botanical garden.

In order to carry out the plan that I had devised, I took a lodging in Santa Theresa, a very pleasant place, near Rio, which was admirably situated to be within reach of the establishments that I had to visit. A small garden annexed to the house was to provide me with the means to deposit the tea plants I would have obtained and to make seedlings.

During November, despite some slight indispositions caused by the climate in Brazil, I traveled relentlessly all around the city of Rio, within a radius of two to three myriameters.[1] I mainly focused my travels towards the mountains of Tijuka and Gavia, whose charming valleys are filled with beautiful houses where, along with coffee, which is the most important product, all the useful plants of the equatorial countries are cultivated.

On November 15, I observed the picking of tea leaves, carried out by black slaves, generally women and children. They carefully removed the tender, pale green leaves, using their fingernails to cut the young leaf bud slightly below the first or second developed leaf.

An entire field had already undergone the operation, and had only tea plants with absolutely no leaves. The overseer tells me that the plant does not suffer at all, and that the harvest of the leaves can become permanent, by regulating it so that the leaves have regrown on the oldest stripped plants by the time the defoliation of the last plants is completed. Approximately twelve thousand tea plants are grown in the garden; they are arranged

---

[1] This is a kilometer, 1,000 meters, or 0.62 miles

regularly in staggered rows, at a distance of about one meter from each other. Most of them have a frail and stunted appearance, which is probably due to the poor exposure of the land, which is at sea level, struck directly by the rays of a blazing sun, and perhaps also to the quality of the soil, whose composition nevertheless seemed to me to resemble that of the rest of the province of Rio de Janeiro. This very clayey soil, strongly colored by iron trioxide, is formed by the decomposition of gneiss and granite rocks. I have often observed all the phases of the transformation of the rock into topsoil; but according to the inclination of the ground, and perhaps also by the effect of other circumstances which I have not been able to appreciate, the soil becomes more or less clayey, sandy, or ferruginous. Thus, when the rainwater has strongly washed the ground and has been able to drain quickly, as happens on very steep slopes, the powdery clay form, is washed away first, then very fragmented quartz, or sand, dominates in the ground, which nevertheless is not completely sterile; for underneath this sand there is a deep layer of clay soil mixed with humus from the detritus of organic bodies. This is not the case of the botanical garden's soil: it retains on its surface a lot of clay and iron oxide, which makes it very compact, and comparable to the strong soils of France.

On the evening of November 16, I met an Austrian diplomat at the home of Baron Rouen, who gave me new information about the cultivation of tea in Brazil, and urged me to make a trip to São Paulo, where there are large crops of this shrub. As a result of this information, I had the honor of sending you a summary of my work on November 17, 1838, and to inform you of my subsequent projects.

Taking advantage of the good will of Mr. Brandao, director of the botanical garden, I accepted the appointment he gave me for November 20, in order to see all the operations of tea preparation carried out there.

The leaves had been picked in the morning, and there were two kilograms of them that were still covered with dew moisture. They were placed in a well-polished iron vase, in the shape of a large, very flared terrine, and placed on a masonry stove. The fire, lit and maintained with wood, brought the heat of the vase to a temperature close to that of boiling water. A negro, after washing his hands, stirred the tea continuously in all directions until the outside moisture had completely dissipated and the leaf had acquired a suppleness similar to that of a linen cloth, or a pinch of leaves rolled in the palm of the hand would form a solid ball. In this state, the mass of tea was divided into two portions, which two negroes placed on a rack made of flat, sharp-edged bamboo strips. They kneaded and stirred the tea for a quarter of an hour, an operation that requires a certain amount of practice, and which has a great influence on the beauty of the products. It cannot be described, because the movements of the hand are very fast, very irregular, and the pressure exerted by the person rolling is a matter of tact which varies from one individual to another; in general, young negroes are more apt to this manipulation than older negroes.

At the same time as the leaves were twisted and rolled, the greenish juice of the leaves was expressed through the screen. The tea must be carefully purged of this pungent and even, it is said, corrosive juice; the purpose of kneading masses of leaves is therefore as much to drive out this juice completely by breaking the parenchyma of the leaf (which would not happen by the rolling of the leaves one at a time), as to encourage the tea to roll up and accelerate the work.

The masses of leaves being rolled in this way, they were put back into the large iron terrine until the temperature had risen so high that the hand could not withstand the heat at the bottom of the vase. The negroes agitated it constantly, separating the leaves, lifting and flipping them to multiply the surfaces, until they were perfectly dry. This work lasted almost an hour, and the manipulators had to be very careful, firstly to avoid burning their hands, and secondly to prevent the leaves from sticking to the bottom of the vase by the effect of the great heat that an uncontrolled fire gives off. — These disadvantages would be overcome by using a water bath, either with pure water or salt water, to increase the temperature of the terrine if necessary, and by using a suitable iron spatula to divide the masses of leaves, stir them, and make them flutter. In spite of the problem I have just pointed out, the rolling and desiccation of the leaves was a complete success: they became increasingly curled, and they retained their rolled shape, except for a few that were too old and consequently too tough to be rolled.

The tea was then put on a sieve or a strainer with large, regular openings (about 3 square millimeters) which was formed from flat strips of bamboo. The best-rolled leaves, from the softest tips of the buds, passed through the sieve. They were winnowed to separate the fragments of unrolled leaves that could have passed through at the same time, and the result is what we call *Imperial tea* or *Uchim tea*. This tea was put back into the vat until it acquired a leaden color which attested to its complete desiccation. It was winnowed, and defective leaves that had escaped winnowing and screening were separated by hand. The tea which was left over after this first sifting was reheated, winnowed, and sifted; it provided *fine Hyson* or *commercial tea*. The same operation was repeated, so that the leftovers from fine Hyson tea became common *Hyson tea*, and then, for the residue of the latter, coarse Hyson tea. Finally, the broken and unrolled leaves, which are the residue of the last winnowing, provide what is known as *family teas*, one of which is of superior quality and bears the name *Chuto*, while the inferior variety is known as *Chato*. The latter types are not sold commercially, and are instead consumed within the tea growing families. Mr. Brandao has been kind enough to give me all the products of these operations, of which I have brought back samples for you to see.

This is the method of preparation used in Rio de Janeiro. — I would add that since the Botanical Garden is to serve as a model for the cultivation and preparation of teas, more care is taken in this establishment than by the private growers, and therefore Brazilian tea in general should not be judged by that of Rio's garden. I was assured that in São Paulo, each grower has his own particular method, which necessarily influences the quality of the products. It was therefore necessary for me to visit this province, in order to achieve the essential goal of my mission, that of knowing exactly the state of the cultivation and manufacture of tea, considered as a commercial commodity.

Steps to Obtain Tea Plants and Seeds

As soon as I arrived in Brazil, I did not lose sight for a moment of the part of your instructions that prescribed that I should obtain tea plants and seeds in abundance. But, based on what I had observed in the garden in Rio de Janeiro, it seemed to me that it was very difficult to obtain plants suitable for export. Nearly all the plants grown there were far too large to be transplanted, and would have taken up too much space in the crates, so I had to reduce the number considerably. Everything I heard about the crops in São Paulo soon reassured me, and I had reason to hope that I would bring back from this country plants in perfect condition for transplanting, and in sufficient numbers to make a sufficient shipment. However, fearing that I would lack this resource, I secured Mr. Brandao's good will to obtain tea seedlings that were easy to separate from their mother plants, although the acquisition of such *material* was very fortunate; at the same time, I asked him for a large quantity of recently harvested seeds, to be sown in our small garden in Santa Teresa, Rio de Janeiro. Mr. Brandao granted me this request with the best grace in the world, and on December 12 I asked him to fulfill his promise. I brought back from the botanical garden about a thousand healthy and perfectly ripe seeds, which can be easily recognized by the brown-violet color of their outer shell. Mr. Houlet immediately set about preparing the ground for sowing these seeds. As the soil was excessively clayey and hard, it had to be broken with a spade, loosened, and smoked; basically, we neglected no care to ensure the success of these seeds.

Research on the Origin and the Determination of Wood Dyes, Medicinal Barks and Drugs — Acquisition of Wood for Construction, Woodworking, etc.

The month of December was excessively hot and rainy. I didn't let a single day of good weather pass without visiting the country houses around Rio de Janeiro, where there was something to learn about teas or plant products that could be of interest to trade and industry. As I wandered through the magnificent virgin forests that provided me with so many beautiful plants for the ornament of the greenhouses of the Natural History Museum and for the studies of botanists, I discovered the origin of a great number of precious woods for dyeing and woodworking, as well as a host of substances used in medicine. Thus, I ensured myself, by collecting samples of wood with its leaves, flowers and fruits, of the botanical determination of the woods known as *Palissandre* or *Jacaranda, Gonzalo-Alvez,*

Vinhatico, and a host of others which are now of such importance that our ships from Le Havre and Bordeaux are bringing back considerable loads of them.

It was quite peculiar, I would even say shameful for science, that these trees, so eminently useful, were less known than most plants which are of no use to society, and almost of no interest to scientific research. The origin of some dye woods, at the top of which I will cite the famous Pernambuco wood, was still a subject of dispute between naturalists, and the solution to this issue has not been indifferent to trade since some traders, whose names I do not think it is prudent to disclose, had compromised their fortunes by speculating on this wood which, in their botanical ignorance, they believed to be supplied by another tree of the same family and almost absolutely similar to the real Pernambuco wood, which the Brazilian government has monopolized. — Thanks to the documents I have collected, either on living nature or from the educated people of Rio de Janeiro, among whom I shall mention Doctors Riedel, Ildefonso Gomes, Sigaud, and Cuissart, and Mr. Soulié, a pharmacist, I have also been able to definitively establish the origin of various barks with very energetic medicinal properties, such as those known as *Pao-Pereira*, *Casca da Anta*, *Fedegoso*, *Parahyba*, *Paratudo*, etc., which have been used in the manufacture of medicines. I brought back most of these products, samples of which I gave to Professor

Guibourt, to be preserved in the collections of the School of Pharmacy.

In my excursions, I often had the opportunity to observe the extraction of real Copahu balm, which is collected by means of wide cuts made in the trunks of the *Copaifera*, very tall trees that are scattered in the forests of the mountains around Rio. I also collected pieces of Copal resin on the trunks and at the foot of the *Hymenœa Courbaril*. Mr. Riedel pointed out to me a genuine species of *Cinchona*, growing in the Tijuca mountains and probably providing a cinchona bark as antipyretic as the cinchonas of Peru, judging by the botanical similarity between this tree and those produced by the latter. In order to ascertain this important point of medical and commercial interest, I have deposited at the Natural History Museum in Paris, a flowering sample of this *Cinchona* which was collected by Mr. Riedel, as well as numerous samples of the tree whose bark is known by the very unsuitable name of Rio Cinchona and which belongs to a distinct genus of *Cinchona*. Finally, to put an end to the report on my findings, I will simply add that I have frequently come across Mr. Auguste Saint-Hilaire's *Ilex Paraguariensis* in the vicinity of Rio de Janeiro, a shrub identical to the one the Jesuits planted in staggered rows in the missions of Paraguay, and whose leaves are a considerable object of trade throughout Spanish Southern America, under the name of *Paraguay tea*. I brought back a live plant of this shrub which I deposited in the King's Garden, along with a species of *Vanilla* and a large quantity of rare or interesting plants.

I used the rainy days of December to study and identify, with the assistance of Mr. Riedel, a huge collection of dry plants and various plant products, which I acquired from a Danish naturalist (Mr. Claussen) who had spent two years on the banks of the Rio San Francisco. To this collection was added another of the utmost importance, which consisted of timber for construction, woodworking, and dyeing, with samples provided with leaves, flowers, and fruits for their botanical determination. These collections are now part of the collections of the Natural History Museum.

Sowing of Tea

In early January 1839, Monsieur Houlet began to sow tea once again, not only in the open ground in our little garden, but also in pans, to facilitate the lifting of the young plants and their transplantation into the cases. The heat being excessive, we purchased mats to protect them from the burning sun, and we watered them more frequently. Many of the seeds we had sown a month earlier had already sprouted but, as the soil was too compact, not all of them were able to break through. Hence, we realized that a lighter soil would be required.

Voyage in the Province of São Paulo — Visits to the Proprietors — Tea Growers — Cultivation and Preparation of Tea in the Province — Costs of Cultivation and Production

It was during this period that, at the invitation of Baron Rouen and Count Ney, I was to go to Serra dos Oragos to visit the tea plantations of Monsieur March. However, I was prevented from doing so due to an illness, which fortunately did not have serious consequences. Thus, I postponed this part of my voyage until my return from São Paulo, which I had set for January 15. Needing the assistance of Monsieur Houlet to care for the tea plants, which I hoped to obtain from the cultivators of São Paulo, I urged him to make the voyage with me, although our seedlings in Rio de Janiero demanded his presence in this latter city. However, Monsieur Pissis, a French engineer and geologist, with whom I had formed an intimate acquaintance, most obligingly offered to attend to the seedlings and our collections, so that I would have no concerns in this regard. Many influential persons in the capital of Brazil hastened to give me letters of introduction to the proprietors and tea growers of São Paulo however, in this respect, I am the most indebted to Doctors Cuissart and Sigaud. The family of Monsieur Venancio Lisboa wrote favorably on my behalf to the governor of the province who, as I mentioned previously, is one of their relations. Monsieur Riedel sketched out a very detailed route for me and included observations worthy of note along the way. Finally, Monsieur Ildefonso Gomes gave me a touching example of his benevolent friendship, taking time from his numerous patients and offering his services to introduce me to the people to whom I had been recommended and to act as an interpreter.

We left on the 15th of January, by steamboat and in two days, reached Santos, the principal port in the province of São Paulo. We crossed the great chain of mountains named the Serra do mar, in caravans drawn by mules, and arrived in the city of São Paulo on the 19th of January. The following day, Monsieur Ildefonso Gomes, accompanied by his relation, Monsieur d'Azembuja, a lawyer who had studied in Paris, introduced me to Messieurs V. Lisboa, governor of the province, Raphael d'Araujo Ribeiro, brother of the Brazilian Minister in Paris, and to Major da Luz, a wealthy landowner of the country. We were very warmly received by these gentlemen, thanks to the kind recommendation of my patrons who made them understand that my mission was in no way prejudicial to their interests, and that it was advisable to show friendship towards the French nation, which had always been so considerate towards foreigners, and Brazilians in particular. We then paid a visit to Monsieur Vergueiro and Monsieur da Costa Carvalho, former Regent of the Empire, for whom Monsieur Sigaud and Monsieur Cuissart had given me letters, but they were not in São Paulo at the time. Foreseeing that my stay could be prolonged until the middle of February, I took lodging in the city's only hotel, operated by a Frenchman who treated me with all the respect due to a fellow-countryman.

Monsieur Ildefonso Gomes and Monsieur Barandier, a historical painter whose desire to see new peoples had induced him to become my traveling companion, wanted to accompany me on the 21st of January during my visit to Monsieur Feijo, former Regent of the Empire and now President of the Senate. We found this venerable ecclesiastic at his country house, situated two leagues from the city. He took us immediately to his tea plantations and showed us all the operations for the preparation of the leaf. About three kilograms of fresh leaves from the previous evening's harvest were used. The tender, flexible and not brittle leaves had been gathered along with the petiole and the tips of every bud. A negro placed the tea in a well-polished iron pan, made in the United States, like the ones I had seen in the garden in Rio, that is to say, about one meter in diameter and thirty centimeters deep. He added a little water, probably to prevent the leaf from burning, as the pan had been overheated. When the leaves were sufficiently *dry*, three negresses rolled them on a bamboo sieve, in exactly the same manner as in Rio, with the only difference that they kneaded their heaps of leaves more skillfully and in all directions. After the leaves were kneaded, twisted and rolled, they squeezed them between their hands and expressed the juice into a jar. The heaps of leaves were returned to the pan and a negro dried them by constantly shaking them by hand. This shaking process was not fast enough, and the fire was not properly controlled to prevent the tea from adhering, at least in part, to the sides of the pan. This lack of attention which would certainly adversely affect the quality of the tea. Finally, when the tea was perfectly dry, it was removed from the pan and placed in a box to protect it from the air and light. — Monsieur Feijo informed me that this was the entire process commonly practiced in Saint-Paul but that, when the cases needed to be hermetically sealed, the tea underwent an additional desiccation process over the fire.

The plantations belonging to Monsieur Feijo, which surround his chagara,[2] are quite extensive. They contain about twenty thousand plants which I judged to be very vigorous and of various ages, most of them from six to eight years old. - These shrubs were planted in regular rows, about a meter and a half from each other with a meter and a half between the rows. The soil is excellent, argillaceous-ferruginous, as is generally the case in the vicinity of São Paulo. — Mr. Feijo then took me to other parts of his property, where I noticed a complete assortment of European plows and other agricultural implements. — That same evening, I visited the Botanic Garden of the city of São Paulo, where some square parcels of land are devoted to the cultivation of tea, but I do not know whether the leaves undergo the preparation process.

Major da Luz had invited me to visit his tea plantations near Nossa Senhora da Penha. I went there on the 27th of January, accompanied by Messieurs Barandier and Houlet. The crops of this landowner were exceedingly well-kept. The ground, which was almost level, was previously flooded and had been drained through the immense labor of Monsieur da Luz. The nature of the soil is less argillaceous than in other places, and the large amount of vegetable detritus remaining in

---

[2] Country House surrounded by gardens and orchards

the fallow land gave it the appearance of richly manured soil. Furthermore, the tea plants have a vigor that I have not observed elsewhere and almost all of them have attained the height of two to three meters. They are well-aligned and distanced from each other so that a man can easily go all around each plant. Below the tallest shrubs, we were pleased to notice many seedlings that had grown naturally from fallen seeds. We asked Monsieur da Luz for permission to take as many of these seedlings as we could manage. Unfortunately, a large portion of the plantation had been recently cleared. We were unable to observe the preparation of the leaves in this fazenda[3], because it was a Sunday. Monsieur da Luz informed me that, although the harvest proceeds without hardly any interruption throughout the year, it is however more substantial during August, September and October, which are the spring months in the province of São Paulo. When I bid farewell to Monsieur da Luz, he showed me his tea warehouses which were quite substantial.

Monsieur Raphaël d'Araujo Ribeiro had made me promise to spend a few days at the property of his mother-in-law, Dona Gertrude Galvao e Lacerda, located at the foot of the Jaragua, a mountain famed for its gold mines. I went there on the 28th of January, in

---

[3] This is the name of an extensive rural plantation.

the company of several Europeans and Brazilians of distinction. I spent two days exploring this celebrated locality, where I was overwhelmed by the kind hospitality of Dr. Raphael and his wonderful family. Not only did they provide me with transport mules for the journey, but they also gave me letters to present to their farmers in Bras and Ypiranga, asking them to provide me with mules and servants, show me all the aspects of tea cultivation, and allow me to take any young plants that I desired. On my way back from Jaragua to São Paulo, I went to see Colonel Anastasio, whose fazenda is at the Tiété bridge. According to Messieurs d'Andrada (Martin-Francisco and Antonio Carlos), this venerable grower has, at the present time, the most beautiful crops and the best tea production. I was graciously received and he himself took me to see his plantations. They are in the most prosperous condition, situated behind the habitations on a gentle slope and on a well-manured tract of land. The shrubs are generally kept small, to make it easier to harvest the leaves. They had been carefully pruned by cutting off the upper branches to keep them low and force the shrub to branch out. When the leaves are picked, a certain number of buds are left on each branch. There are about fifty to sixty thousand plants in this establishment, but of this number, a third of them were planted just the year before. Colonel Anastasio then took me to the premises where the tea is prepared.

Everything here is in good condition and in perfect order. The pans are made of iron, like those of Monsieur Feijo, and the ovens are also low and with flat edges. With prolonged use of these pans, the iron has somewhat deteriorated, probably due to frequent exposure to the fierce heat of the fire. A saltwater bain-marie would prevent this by keeping the temperature at a constant heat. The *taquara* (bamboo) screens and mats are quite elegantly made. In general, there is a cleanliness that prevails in this establishment which is pleasant to observe, and which gives a favorable impression of these products. The colonel showed us the warehouse where the tea is stored. There were many pot-belly tin jars, with a conical upper portion tapering to a narrow-neck closed by a tight-fitting stopper. These vessels, which resemble those of our Parisian dairy farmers, contain from half an arroba (about 8 kilograms [17.5 pounds]) to three arrobas (48 kilograms [106 pounds]). If they are all full, I would presume that Colonel Anastasio has one hundred arrobas of tea in his warehouse. Using the utmost discretion, I posed a few questions in this regard and was told that he is by no means in a hurry to place his teas on the market. He is waiting until it is requested from him, hoping that, by keeping the tea in the warehouse longer, it will only improve with time and, by limiting the supply, its value will not depreciate. I then spoke with him about the cost of cultivating and producing tea in Brazil. He said that it was so great that, in order to make a reasonable profit from growing tea, it had to be sold for a price of at least 2,000 reis (about six francs) a pound. In fact, the work in Brazil is done entirely by slaves who, in truth, do not cost much to feed, but work as little as possible, as they have no interest in it. Moreover, these slaves command a high price and the chances of mortality, along with the exorbitant rate of interest on money, further increase their market value.

On the 31st of January, Major da Luz kindly presented me with about three hundred tea plants, which he had asked his negroes to pull up for me. On the same day, I went to Ypiranga, a farm belonging to Dona Gertrude Galvao, where there was an immense tract of land planted with tea that had been neglected or, in any case, was not considered to be an object of value to the proprietor. Monsieur Raphael d'Araujo Ribeiro had given me a letter for the overseer of this fazenda, and I made arrangements with him to help me remove all the seedlings that I wanted. Indeed, I returned there on the 4th of February and, in a single day, Monsieur Houlet and I, with the help of some negroes from the farm, managed to pull up about *three thousand seedlings* which we carefully arranged in large bamboo baskets (*cestos*). To lighten the weight of these shrubs, Monsieur Houlet left as little soil as possible around their roots, but he was careful to wet this soil well before closing the baskets, which he covered with banana leaves.

Despite my interest in studying tea cultivation in São Paulo, I felt that there was no need to extend my stay in this country any longer. All that remained for me to visit was the establishment of the late Monsieur Jose Arouche de Toledo Rendon. I owed my introduction to this beautiful property to Monsieur Clémente Falcao, a professor at the law school, whom I had met at the home of the President of the province. Since the death of Monsieur Arouche, the chacara has not been as well-kept, although it is still the largest that I have seen devoted to the cultivation of tea plants. They are set in rows quite far apart; the plants are beautiful and tall, forming strongly branched shrubs from one meter to one and a half meters high. The spaces between the shrubs have been planted with maize. The borders of the square parcels of this vast plantation of more than fifty hectares, and surrounded by beautiful alleys of *araucaria brasiliensis*, are composed of little dwarf tea plants which are kept low by cutting their main shoots down to the level of the soil. Only eight negroes are employed for gathering the leaves; each negro can pick about a half-arroba (8 kilograms) per day.

Departure from Saint-Paul — Voyage to Ubatuba — Information on the Trade of This Country

I left São Paulo on the morning of the 6th of February. I stopped for a while in Ypiranga to have our tea loaded on mules that would transport it to Santos, through the great Cordillera or Serra do Mar.

On the 8th of February, I embarked on a steamboat bound for Rio de Janeiro. When we came within sight of San Sebastian, I instructed Monsieur Houlet to proceed to the city alone and charged him with taking the greatest care of our tea plants upon his arrival in Rio, while I would visit the flourishing colony of Ubatuba, inhabited by French families who very successfully cultivate coffee and other useful vegetables. One of the main growers of this country, Monsieur Vigneron de la Jousselandière, owner of a tea plantation, had urged me to come and visit him. He had instructed me on how to reach him through Captain Jacintho de Godoi, who provided me with a pirogue and some rowers. After a delightful sail through an archipelago of enchanting islands, I landed in Pontagrossa, near Ubatuba, where I was wonderfully received by Monsieur Vigneron, Emile and Camille Jan, Robillard, Richer, Gauthier and other French planters. During the eight days I spent in their company, I obtained valuable information, not only on the colony's cultivation and trade of tea, but also on the trees that grow naturally in the virgin forests of this beautiful land, and which provide useful wood for building, cabinetmaking and dyeing. I brought back authentic wood pieces from several species and, under the dictation of Messieuers Vigneron and Jan, I drew up a complete list of those that I could not obtain, due to lack of time to collect them. — My stay in Ubatuba allowed me the opportunity to copy a hydrographic map of the points of the coast that Admiral Roussin had been unable to identify, as well as the plan of the excellent port of Ubatuba. I had this document dispatched to the Depository of Navigational Charts through Monsieur Cecille, the Captain of the vessel. Finally, I visited the tea plantations of Monsieur Vigneron, which are very beautiful, but which the owner is not cultivating to their full potential, preferring to dedicate himself to growing coffee, which is very lucrative. He permitted me to take away a large quantity of young tea plants, as well as cacao trees. It was with deep regret that I took leave of these worthy settlers and, on the 18th of February, I re-embarked on a Brazilian schooner which took me to Rio de Janeiro in a few days.

Return to Rio de Janeiro — Fabrication of Cases for Tea Plants

Upon my arrival, I found the tea plants from São Paulo, set by Monsieur Houlet in our garden in Santa Theresa. To these plants, I added the stock that I had brought from Ubatuba. The youngest seedlings had almost all perished along the way due to the excessive heat. Mr. Houlet had those which survived covered with mats and gave them a great deal of care to revive them.

It was at that time that I sent you, Honorable Minister, a letter dated the 27th of February, in which I outlined my operations in São Paulo and Ubatuba. However, I was beginning to have some concerns regarding my return to France. The news from La Plata did not augur well for a forthcoming lifting of the blockade and, although the details of our glorious expedition against Mexico[4] were known in Rio, false

---

[4] Guillemin is referring to the First French Intervention in Mexico or "The Pastry War," a French blockade of Mexican ports from November 1838 to March 1839 set up in response to claims of financial losses by French nationals in Mexico.

rumors were being spread in an effort to diminish these results and prolong the resistance of Buenos Ayres. Nevertheless, in the hope that French vessels from La Plata would reach Rio in April, and that I would be able to load my collections there, I determined that it was time to turn my attention to the fabrication of cases in which I could pack them. As I was unable to come to an arrangement with French or Brazilian carpenters because they demanded an exorbitant sum for their work, I decided to purchase the necessary wood and iron myself and hired two negro carpenters, by the day, to construct the cases under my direction and the supervision of Monsieur Houlet. My first intention was to construct boxes using the Wardian system, consisting of airtight cases with stained stained-glass windows on top to let the light in, but I was forced to abandon this plan because of the high cost. However, due to my success at sea with a chest containing fruit trees, which I had given to the Consul General in Buenos Ayres, I decided that a case which could be closed or opened at will by means of movable panels would be perfectly suitable for tea plants which, moreover, need to be checked and watered from time to time. The construction of these cases, as well as those intended to contain the hothouse plants for the Museum, caused me a great deal of trouble and worry. However, this did not prevent us from pursuing our research and studies, at the same time, in the vicinity of Rio de Janeiro, during the months of March and April.

As Monsieur Dumas of the Academy of Sciences had asked me to procure samples of a vegetable wax from Brazil in which he had found a new ingredient, as well as information on the origin of this substance, I took it upon myself to fully satisfy this request. I sent him a specimen of Carnauba, a substance that is somewhere between resin and wax, and which the inhabitants of northern Brazil use as an article of trade with Buenos Ayres, Montevideo and even England. Dr. Sigaud provided me with everything that had been published in Brazil on this substance, and on other vegetable waxes from South America. I purchased a book that is eminently useful for the trade and agriculture of Brazil, by Messieurs Taunay and Riedel[5] and recently updated in a second edition, which contains a table of Brazilian roots, bark, wood, fruits, gums, etc., with their vulgar and scientific names, and brief information on their uses. This publication also contains cultivation records for coffee, tea, cotton, and other colonial products.

---

[5] C.A. Tauney and L. Riedel, *Agricultor Brazileiro Memoria, segunda* edição (Typographia Imperial e Constitucional de J Villeneuve e Comp. Rio de Janeiro, 1839)

Preparations for the Return to France on the Corvette Heroine — Tea Plants and Seeds obtained from the Botanic Garden — Care of the Tea Cases by Captain Cecille — Effect of Crossing on the Tea Plants

As I continued to be concerned about returning to France at a good time, that is to say, before the bad season, I wrote on the 22nd of March to inform you that if, during the month of May, a warship did not arrive on which I could voyage, I would return to Toulon or Marseilles by merchant ship. Fortunately, on the 9th of May, the *Heroine*, commanded by Captain Cecille, arrived in Rio, rendering this precautionary plan unnecessary. At that time, I was in Serra dos Orgaos visiting the large agricultural establishments of Monsieur March, where I hoped to obtain more tea plants. I returned to Rio with great haste. Straight away, I saw Baron Rouen and, through his kind intervention, it was decided that Commandant Cecille would take me on board his ship with my collections. Unfortunately, our cases were not yet completely finished, and I needed to visit to the Botanic Garden once more to procure tea plants and the greatest possible quantity of seeds. Monsieur Cecille immediately provided me with two ship carpenters who, in just a few days, did more work than my negroes had done in a month. He also sent me several sailors to help carry the cases from the top of Santa Teresa to the place of embarkation. On the 16th of May, I paid my last visit to the Botanic Garden where I procured seven *hundred* young tea plants, taking care to choose those with roots and a height of three to four decimeters. To these I added at least *two thousand* perfectly ripe seeds, sown by Monsieur Houlet in the spaces between young plants numbering about three thousand, the whole of occupied the eighteen cases for the Ministry of Commerce. Finally, all my preparations being complete on the 26th of May, I embarked that same evening, after bidding a reluctant farewell to the many friends that I had made in Rio de Janiero.

It was a very pleasant sight for me when, the day after the departure of the *Heroine*, I beheld the eighteen cases of tea that had been perfectly stowed by the commander with great care. They had been placed in the battery deck, arranged in twos, back-to-back, between the pieces of artillery, except for a pair that had been placed in the middle of the battery deck. This allowed them to receive light and the panels could be easily closed in case of bad weather. The vigor of the tea plants, the beautiful verdure of their foliage, in a word, the prosperous state of these plants that we had admired in Rio, lead me to anticipate the most successful outcome from my expedition. No longer did I regret not arriving in France by way of Toulon, since I could see the great interest that Monsieur Cecille took in my beloved plants. What other naval officer would offer such protection and care to my plants? Furthermore, it was not imperative for my tea plants to arrive directly in the South of France, since several localities in the west and central part of the country could, in my opinion, claim their share. Therefore, at the time of my embarkation, everything seemed to be proceeding according to my wishes. But this moment of satisfaction was very short-lived, for on the 29th of May, the winds blew us south out of the tropics, the sea became rougher than is common in these parts, and we were forced to keep the gun ports closed, lest the sea water should cause irreparable damage to our plants. The absence of light during the first days of our navigation was therefore the initial cause of the deterioration of our young tea plants, particularly those that were newly planted. After several days, when the sea became calmer, permitting us to keep the port holes open, the breeze swept across the surface of the waves, casting a very fine spray of salt water on the cases. This undoubtedly resulted in great injury to the plants, since the cases exposed to the wind suffered infinitely more damage than those on the other side. By the 11th of June, at a latitude of 4° 22' south and longitude of 32° 6' west, the tea plants had lost most of their leaves and many of the stalks had dried up. However, I hoped that some would re-sprout from the root. Many seeds had germinated; the young shoots were slender, long, withered and had pale leaves. By the 2nd of July (latitude of 24° 36' north, longitude of 42° 53' west), the strongest tea plants were severely suffering. However, a few had grown offshoots and the germinated plants had taken on a greener hue. During the cannonball exercise conducted that day, Monsieur Cecille secured my cases, to ensure that they would not be affected by the strong blasts. He ordered that I be given fifty liters of water twice a week for our tea plants. I cite this fact as a testament to the Commander's concern for my plants, since, as a result of the leakage of some water casks, the crew had been placed on a strict ration of three quarters of a liter a day.

Arrived at Brest — Difficulties in Transporting Teas to Paris — Agricultural Excursion in the Finistère Department — Care Given on Their Arrival in Paris

The *Heroine* anchored in the harbor of Brest on July 24th, at four o'clock in the evening. The very next day, I went to see Admiral Grivel, Maritime Prefect, who had not yet received any order from the Minister of the Navy to transport my plants to Le Havre. On the same day, I had the honor of announcing my arrival in France, and I begged you, Minister, to get me out of Brest promptly, either by steamer or by any other means. I pointed out to you the dangers posed to our tea plants, which were visibly withering away, and that the prolongation of the journey was very tiring. The only brig of war, the *Salmon*, which the Admiral could place at my disposal, had appeared insufficient to his captain and to Mr. Cécilie for the transport of my collections; I thus made the decision (which you kindly approved) to have my boxes hauled to Morlaix, to then embark them aboard the steamer from Morlaix to Le Havre, and to have boats drawn by horses transport them up the Seine. During my forced stay in Brest, I spent my time visiting the western end of the Finistère department and studying its climate, soil, and crops. Messrs. Crouan, brothers, pharmacists, and Mr. Paugam, gardener at the naval hospital, were kind enough to accompany me to the most important places for observation. I am convinced that nowhere on French territory would cultivating tea be easier than in certain localities in Finistère, where the temperature, quite mild in winter, allows camellias to be grown in the open ground, where the humidity of the atmosphere produces an extraordinary development in the leaves of trees and shrubs, where the soil, similar in nature to that of Brazil, seems to be perfectly suited for tea as well as small crop plants, where finally the labor, among a poor and ignorant population, would be at the cheapest possible price.

Arriving in Paris on August 28th, the Brazilian teas were deposited in the King's Garden. Mr. de Mirbel ordered Mr. Neumann, chief gardener, to prepare greenhouses and layers for the surviving plants, which number about fifteen hundred, including those from germinated seeds. Mr. Houlet continues to care for them, and I have reason to believe that around next spring, they will be fit to be transported to the departments you will have chosen by that time.

After having briefly outlined the events of my trip, Mr. Minister, I will now present you with a summary of its material results and the information I have gathered on Brazilian teas. I will attach some considerations on the advantages and disadvantages, the facilities and difficulties that the cultivation of this shrub and the preparation of its leaves in France may present.

Material Results of the Mission — Tea Plants and Seeds and other Products

I had procured about three thousand seedlings of young teas, which I had placed in eighteen boxes, and between which I had sown more than two thousand ripe seeds. The sea voyage killed more than two-thirds of them, so that there are only twelve to fifteen hundred of them left, which are currently stored in the Natural History Museum. This is precisely the number you indicated in your instructions to me; but I should point out that my relations with Brazil will allow me to bring from that country the quantity of tea plants and seeds that we will want. This would not be a very expensive shipment, as many of the costs I had to incur could be avoided, and we'd arrange it so that the crates would be small. I don't think it is right to bring in plants that are already strong; recovery is too difficult for them, and they take up too much space. On the contrary, seedlings from seeds sown in appropriate crates would have the best chance of success. In the meantime, and in order to use the seedlings I brought, I entrusted them to the care of Mr. Neumann and Mr. Houlet, who separated them into pots, and placed part of them in a temperate greenhouse and part of them in layers to accelerate their vegetation.

I brought back more than 150 species of wood, collected in the provinces of Rio de Janeiro, São Paulo, and Minas Geraes. Most of these woods have uses in lumber, woodworking, and dyeing. Their origin and botanical determination has been made in situ, and can again be ascertained by means of samples with leaves, flowers, or fruits, which relate to these woods. I should also mention, as a result of my voyage, the beautiful collection of woods suitable for construction, collected in New Zealand by Captain Cécille, who generously gave it to me, and which I donated to the Museum of Natural History.

A large number of plant products, consisting of gums, resins, bark, roots, etc., the exact determination of which was very important for the drug trade, were given to Mr. Guibourt, a professor at the School of Pharmacy, who will place them in the collections of this establishment.

The plan of the port of Ubatuba and the coast surrounding this colony, which was of interest for trade in general, and French trade in particular, since it owed its nascent property to the French, was submitted to the Navy's chart office.

Information on the Climate and Soil of Brazil — Tea Cultivation — Sowing — Planting — Harvest Operations Related to the Preparation of Various Kinds of Tea — Color and Fragrance of Tea

In the short space of seven months, I visited almost all the tea-growing establishments that exist in the provinces of Rio and São Paulo. I will give a general overview of the climate and soil of these regions, the cultivation of teas, and the work that goes into it, which does not vary significantly from one establishment to another.

The climate of Brazil seems very suitable for growing tea; I have seen this shrub thrive in the hottest places, such as the Botanical Garden near Lake Freytas[6].

---

[6] Based on Dr. Pissis's observations, the mean temperature at Rio de Janeiro is 23.5° centigrade at sea level.
The temperature of the soil is:
   at the surface A maximum 45° in the sun, minimum 13° during August rains.

If the plots of this garden devoted to tea growing contain only plants of rather stunted appearance, this is due to special causes, perhaps to their exposure too close to the sea and to the nature of the soil, for in the part of the garden which is close to streams of flowing water, and where beautiful nurseries of exotic trees can be seen, very vigorous tea plants can be observed. However, the cooler climate of São Paulo and the Serra dos Orgaos seemed to me to be more favorable to the teas, whose vigor is admirable there. I was assured that in *Ouro-Preto* (Minas Geraes) they were also beautiful. Judging by the European plants that grow successfully in the latter areas, they have similarities in climate to our southern departments. Although I was in São Paulo in the middle of the summer, I was not bothered by the heat; and it felt like I was living in southern Europe. This is an effect not only of the latitude, but also of the greater height of the plateau in this province.

The cultivable soil of Brazil is generally clayey, ferruginous, resulting from the decomposition of granitic gneiss rocks, and more or less penetrated by humus. This land, which has similarities to the strong lands of the departments formerly of Brittany, is perfectly suited to teas. I have seen them cultivated at several different

---

at 2 decimeters below the surface of the soil Amaximum 39°, minimum, 16°
at 1 meter below the surface A23.5°
at 5 meters below the surface A23.5°

exposures, which is made possible by the gentle breezes of these regions, but they seem to be best suited to the sun exposure on the slopes. Does this exposure affect product quality? They either didn't know, or didn't want to tell me. The land is always carefully prepared by ploughing with a hoe, and often by adding a little manure.

Tea plants are easily obtained from the sowings which usually take place in January, February, and March. When the seed is sufficiently ripe, it is sown immediately, or very soon after being harvested, because it quickly loses its germination capacity. Tea plants produce such a quantity of flowers and fruits that many seeds escape them, which germinate under the old plants and are used to feed the plantations. These are usually arranged in a staggered pattern, with the shrubs about one meter apart to make it easier to pick the leaves. Some farmers plant in widely spaced rows, and in the intervals they grow maize or other economic crops.

Although the tea harvest can take place nearly all year long, it is in the months of October, November, December, January, and February that it is most active. Negro slaves, often children, are tasked with picking the leaves using their nails to cut the most tender leaves and the tips of the young buds. The work of these slaves is more or less expeditious; but it is still very expensive, compared to how much the same work would cost in

Europe.[7] I have been assured that a day's work for a negro costs his master about two francs, if one includes in this account, in addition to food and clothing, the interest on the purchase price and the chances of mortality. It is estimated that a good worker can harvest up to seven to eight kilograms of leaves per day. It is a general rule that a tea plant must be three years old before the leaves are harvested; however, this rule must be adjusted according to the vigor of the bushes. Most growers, and this is how I have seen them operate, mix all the leaves of the same crop to make the various kinds of tea, but I have been assured that it is more advantageous for the products to sort these leaves, which is done by laying them out on a table, and separating the softer leaves used to make *Imperial tea* from the harder leaves ones used to make *Hyson tea* and other commercial varieties. The leaves are harvested in the evening and in the morning of the day on which they are prepared for rolling and drying.

Having described in sufficient detail, in the course of the preceding narrative, the operations required to turn tea into a commercial commodity, I will content myself with briefly summarizing these operations, proposing moreover to make the details of cultivation and manufacture more fully known in an unpublished

---

[7] Slavery was abolished in France in 1848, nine years after this work was published.

report by the late Jose Arouche de Toledo Rendon, whose translation I will publish shortly.[8]

The first operation is to cook *the leaves properly*. To do so, they are spread out in a very flared basin or terrine, made of well-polished iron, placed on a masonry stove, under which a very bright fire is lit. The leaves are cooked when they acquire a soft consistency and can be rolled into pellets without deforming. Then the second operation is carried out, in order to express the greenish, bitter juice, and consequently to roll up the leaf. It is performed by kneading the leaves on bamboo mats with wide meshes and sharp edges. These piles of leaves are turned in all directions to make each leaf roll up perfectly. This task lasts approximately half an hour. In a third operation, the leaves are put back over the fire, still in the same basins, and they are stirred continuously with the hand, making them jump and flutter to speed up the drying process. Care must be taken not to let the leaves stick to the bottom of the pan for fear that they will burn and blacken. During the drying process a lot of dust escapes, which is simply the fluffy down coming from the

---

[8] José Arouche de Toledo Rendon (1756-1834) was a military leader, director of the São Paulo Law School and pioneer of Brazilian tea production on his large plantation outside the city of São Paulo. We have not located a translation of Rendon's report by Guillemin; however, a report on tea growing and processing by Rendon is included in *Agricultor Brazileiro Memoria* mentioned earlier by Guillemin. This appears to be the report to which he refers. A translation of that report by Rendon is included as an Addendum to this edition.

fine hairs which naturally cover the leaves. A brush is used to remove this cottony down, which sticks to the edges of the basin. When drying is complete, the fourth operation is carried out, which consists of removing the tea from the basins and screening it in bamboo sieves with holes of various sizes. The first step is to use fairly fine sieves, whose holes all allow the best-rolled leaves to pass through, i.e. those that were the most tender and formed the tips of the young buds. The tea that comes from this first screening is *Imperial tea* or *Uchin tea*. The tea is then winnowed to separate the unrolled leaves, which are then carefully removed. The remainder is placed back in the basins, and lightly heated again. It is sifted through a sieve with a slightly looser mesh than the first one. The product of this sifting and winnowing is fine *Hyson tea*.

The same operation successively repeated results in both *common Hyson tea* and *coarse Hyson tea* for trade. Finally, the last remainder, which is composed of leaves that have refused to roll up, is used to for what are known as Family teas, which can be further distinguished in two varieties, called *Chato* and *Chuto*.

Thus, the various types of Brazilian teas are all obtained from the same harvest, and their differences only come from successive sieving and winnowing.

The leaden grey color of the tea is due to a light roasting that it undergoes before being put into tins, which are kept away from damp air and light.

The tea emits an herbaceous odor immediately after drying, which is not pleasant. After a while, it acquires a special aroma that grows stronger and stronger, and it is only after a year, or even more, that it is good to be used. Brazilians do not flavor their teas because they are unaware of the processes used by the Chinese to flavor their teas. They claim that the smell of good tea is natural, and they condemn the artificial means that are said to be used to flavor the various kinds that come from China. However, they have *Olea fragrans* and *Camellia Sesanqua* in their gardens, which for a long time have been reported as the shrubs which the Chinese use to give a pleasant scent to their teas. *Olea fragrans* flowers emit a delicious scent; I have no doubt that these flowers do play a great role in the flavoring of Chinese teas, although I don't deny that the quality of the tea can depend a lot not only on the shrub itself that provides it, but also on the care taken in the preparation of the leaf. I didn't think it was necessary to bring back *Olea fragrans from Brazil*, because I knew it was grown in France in the gardens of horticulture enthusiasts. As for *Camellia sesanqua*, it is probably by mistake that it has been mentioned as a tea flavoring, as there is nothing aromatic in this shrub. Its resemblance to tea may have led to its leaves being substituted for those of tea on certain occasions.

Results of the Above Observations for Tea Cultivation in France

From my observations of the climate, soil, and exposure of tea crops in Brazil, I can confidently conclude, by analogy, that the warm southern regions of France, Corsica, and Algeria are eminently suitable for the introduction of tea cultivation. Clayey, ferruginous soils and exposure on the slopes of hills will suit it better than light soils and plains. Since it cannot be expected that this crop will be able to expand very considerably, it will be necessary to choose mountainous localities which have hitherto been of little benefit to agriculture, and above all to find out exactly whether the price of labor is not too high there. In my narrative, I spoke of the possibility of attempting tea cultivation in the departments of former Brittany. I stress this idea again, because the humid climate of this country has a very favorable influence on the development of the leaves of several shrubs similar to teas. Thus, camellias spend the winter in the open ground in various localities of the department of Finistère, where they stand out with their beautiful foliage. The same is true of the trees and shrubs of southern Europe and northern America, such as the *Rhododendron*, the *Arbutus*, the *alaternus*, the *fig tree*, the *olive tree* itself (which in truth does not bear fruit there), which we have somehow acclimatized to these regions. The strongest objection that could be made against the cultivation of tea in Brittany, would be drawn from the rainy weather during the short summers of this western region, and which would perhaps not allow for as frequent harvesting as in the south of France to make this cultivation advantageous. Nevertheless, it would be an experiment worth performing, especially if the direction of tea cultivation were to be entrusted to intelligent men of the country, and in this connection, I would particularly draw your attention to Mr. Paugam, gardener of the naval hospital at Brest, who has recommended himself through great agricultural improvements he has made in the country. It would be easy to obtain, from the warships returning from Rio de Janeiro, a large quantity of new plants and tea seeds, using the means I have indicated. These plants could be distributed immediately to farmers in Brittany, and even to those living in the departments bordering the Loire, where tea cultivation has been attempted successfully, but on too small a scale.

Can the cultivation and preparation of tea currently provide an advantageous sector of agricultural industry for France? This serious question, which I am addressing here in anticipation, seems to me still far from being resolved, because the answer is subject to conditions which are very difficult to establish with exactitude, namely: the proportion of products and definitively the cost of the goods. In Brazil, where, as I state in this report, the cultivation of the shrub has been perfectly successful, where the leaves are harvested almost all year round without interruption, where the quality, apart from the aroma (which is believed to be artificially produced), is not inferior to that of the best teas from China, the growers have not made great profits. They produced an immense quantity of tea, judging by the tea I bought in the shops of São Paulo, but they cannot deliver it below 2,000 reis (about 6 fr.) per half kilogram, a price higher than that of Chinese tea of such quality. Trade in the latter commodity is therefore still very active in Rio de Janeiro, either directly from China or indirectly through the United States. But while we should not expect to do better than the Brazilians in terms of quantity of product, we can be sure that the cultivation of the shrub will be as successful in the southern departments and as far north-western France as it is in Brazil; that it will receive improvements which will increase the volume of product; that the reduction in the price of labor in several localities will also bring down the cost price of tea; that the manufacturing costs may also be considerably reduced through the adoption of more economical methods; and that, if we succeed in giving French tea the flavor which so eminently distinguishes Chinese tea, there can be no doubt that it will sustain advantageous competition with the latter, especially if there should be a war at sea or some other disturbance in the present state of commercial affairs, against which we must always be on guard even in the midst of the most enduring apparent tranquility. Whatever future events may occur, the cultivation of tea must be attempted in France with circumspection, and as it will not cause any harm to other agricultural operations, because it will always be restricted to specific localities which are not very favorable to other productions, I think that it deserves continued encouragement and favor from the government.

THE END

Addendum
Additional Reports on
Tea Production in Brazil

Brief Report on the Planting Cultivation of Tea by José Arouche de Toledo Rendon

Tea growing and processing methods in use in Brazil at the time of Dr. Guillemin's visits were recorded by José Arouche de Toledo Rendon. Arouche's report is included in the second edition of the book Agricultor Brazileiro Memoria *that Guillemin brought back to France.*

Brief Report on the planting and cultivation of tea, its preparation until it is ready to be sold offered to the Society in Defense of Liberty and National Independence of Villa de Valença, by its honorary partner José Arouche de Toledo Rendon, Lieutenant General and Director of the Law School of São Paulo.

Everything that has been written about tea only gives a confused idea about it: the practices reported by travelers who have never entered the interior of China, nor seen the factories with their eyes, are not accurate; they are acquired knowledge, from what they have heard from the Chinese who are skilled in deception, in order to preserve the monopoly of a plant of which Europe and America make a general use.

Therefore, the only relevant text we have is the pamphlet by Friar Leandro of Sacramento.[9] This most educated citizen, in addition to his knowledge of botany and agriculture, was the person who skillfully persuaded a Chinese Tea Master to reveal some information he was reluctant to disclose, although he was in Brazil under contract, he stood by the notion that tea was an asset of his nation. I met him privately, and Friar Leandro later, in a confidential letter, expressed the same to me. Such is the strength of love for the country that, even in the soul of the lower classes, it manifests an affection for their birthplace!

Just the information of the pamphlet written by Friar Leandro, about the temperature that the oven must be at during the initial phase of cooking the leaves, and continuing until they exhale their aroma of tea, would have been enough. I, who had not seen this process and was only privy to the literature of travelers who, deceived by the Chinese who only recommended using low temperature, found I was not making tea properly,

---

[9] Friar Leandro do Santíssimo Sacramento (1778-1829) was a professor of Botany at the *Academia Médico-Cirúrgica* in Rio de Janeiro and the first director of the Botanical Gardens in the same city. The pamphlet cited is *Memoria Economica Sobre a Plantação, Cultura e Prepara do Chá* (1825). An English translation is presented after this report.

because using low temperature, instead of cooking the tea and making it manageable to roll, it caused the leaves to fall apart and, when rolled, and reduced them to fragments, like the Congonha herb, called Paraguayan tea. By testing various methods, I discovered the secret; and if Father Leandro had published his pamphlet a little earlier, it would have saved me a lot of trouble.

As a result of what I have just said, it seems that there is nothing more to add in order to make tea; because its point of perfection will depend on the skill of each person, and its improvement will be attained over time. However, I, in this brief summary that I am writing, will contribute something positive to my countrymen. My aim is to show how I plant tea and manufacture it, differing at times from what Friar Leandro wrote. I believe I have simplified some articles, so that our farmers will understand them faster and easier, producing tea with less difficulty, and thus more quickly manufacture it. All reduction in labor increases the earnings of the farmer and the manufacturer.

Tea can be sown in any month of the year in temperate countries; but, bearing in mind that the months of December, January and February are the ones that produce the most abundance of good seeds, and that, therefore, due to the heat and humidity there is more stimulation for the seed to develop and germinate and grow. For this reason this period is better for seed collection and creating the nurseries that serve as deposits for the farmer who wants to transplant the seedlings at any given time. I do not approve the method of planting seeds in the same location where they will grow. They will linger in the land for two and three months; during that time weeds will also grow, a lot of patience and care is required for weeding, abilities that are not innate to our slaves, who will dig up both the invading weeds and the seedlings at the same time with the hoe. Therefore, I prefer the nursery, and I do it as follows:

Having prepared an area with some manure, the earth is well plowed and clods well broken, I divide it into parallel lines, five palms (about 40") apart from each other; I use a string, and I leave the line marked with small stakes that show the course. A person then carefully digs a small trench half a palm (about 4") deep along the length of each line, taking care not to pull out the little stakes that become like markers of the same line. Then the seeds are sown in all of the trenches, either shelled or whole, and in such quantity that they are almost touching. These seeds are covered with the soil from the trench, and, in addition, more soil is pulled from each side, so that on top of the trench, in the direction of the stakes, forms a mound measuring half a palm in height. Care must be taken to ensure that the soil is moist when the seed is sown, and, if not, the trench must first be well watered, because though that topping is made to keep the soil moisture close to the seeds, despite the sun, it also prevents the humidity of sparse rain showers from penetrating it. Once the nursery is done in this manner, the only thing left to do is to repeatedly clean out the weeds, so that it is easy to recognize the spiky stem of the seedling, which is stiff and purplish.

Whoever has an abundance of seeds and plants in large numbers, does not need to take care in choosing them; for when many seeds are planted, the good ones sprout, and the damaged or altered ones serve as nutrients for the others.

Once the planting is done in this manner, the new plants are born in clusters along the rows. At the end of a year or less, they are a palm tall, and they are ready to be moved to their planting places. It is prudent not to let them grow fully, because their taproot is longer than a branch; and if this root is cut or poorly placed in the hole by the planter, the plant will wilt and dry up. Because of this and for other reasons that cause the death of many seedlings, I retain not only the less developed ones, but also, as a precaution, I have three to four seedlings planted in each planting spot, so if some of them die, others will prosper.

Just one seedling is enough to make a leafy clump of tea; but I practice the opposite [placing multiple seedlings in one spot] for the reason already mentioned and considered; the result is that the planting area does not remain empty, and because there are many seedlings, one of them becomes stronger and takes over the others, other times two and three grow with the same vigor, and it looks like a single plant producing different leaves, as seen in the variety that has appeared in São Paulo, which I didn't see in the Lagoa Garden location. The best way to get them out of the nursery is to use a digger or lever on one side and lift the soil: they all come out with the roots whole, and then a sufficient portion is taken, and they are placed into the holes three and four at a time, then another person plants them, uniting them together, and packing the soil from the tips of the roots up to the surface.

Because my tea plantation is in a small farm where the number of plants in my initial plan cannot fit, since forty-four thousand or so plants already crowd the land space, I have increased the number of paths, which all are bordered with this plant, as well as the divisions of the vegetable and other small plant gardens, thus adding one good thing to another: in the middle of this, I have lots of different extensions, as well as triangular beds and other different shapes to accommodate the spaces accordingly.

In the borders of the paths and of the flower beds, or in the different spaces, I no longer use the method of keeping four palms of distance between one plant from another. My rule is to mark the lines, leaving an interval of eight palms between one plant and the other. These lines are drawn with string from a given point to another point then marked with small stakes. They serve as guides to the slave who makes the trench along the entire line, and for the planter as well. This trench should be a palm or more in depth and the earth that is dug out is placed on the edge all the way to one side. Another worker, if the land, like mine, is not very fertile, will spread manure over the earth; and when done, refill the trench with the same soil, mixed with the manure. When this is finished, the same one that dug the trench or any other worker, digs holes along the entire line, uniting them as much as possible, these holes which are a palm in width and dug with a hoe, have the small plant placed in the middle of the hole, and as a general rule, have a distance of at least two palms from each other.

It may seem that this method is an extravagance of mine; however I apply it for reasons of economy. The first function is the view this method provides, hanging strings that act like small walls; and also seeing the clearings that remain between the lines, in the same way that an infantry corps is formed with open ranks; the second is the ease of harvesting without the harvester being subject to the interference of morning dew and rainy days, and, at the same time, each one being in charge of his row to harvest from one side and then from the other side, and everyone within sight of an overseer who, inspects everything; the third is the ease of weeding and tilling the tea plants, without the danger of the hoes harming the branches, and the weeders also being within sight of the overseer; the fourth use is harvesting corn, which is planted annually in rows through these openings, and it produces well without hampering the tea.

The main management of this plant is to dig the ground well annually; since, in addition to being a well-known rule in agriculture that tilled earth is like a sponge which attracts and receives the vegetation's own particles from the air, it is palpably seen that, at the end of the tea harvest, which lasts from the end of September until around the end of May, the earth is relatively well packed down; and every farmer knows that land, in this state, does not produce. The hoe or plow, in the hand of the skillful farmer, fertilizes the ground. Therefore, when the harvest is over, I only make the following improvement.

Each worker takes his row, weeds two or three palms, digs a cross hole a palm and a half deep, and throws the debris and vegetation that he has weeded in it, and goes on weeding up to four or five palms, and rakes this debris with the hoe filling the hole and covers it with the earth that he continues to dig from the first hole until almost the end of the weeding, he makes another hole fills it with his weeding going forth, and continues digging using this method until the end. Once they get used to this mode of weeding, they take as much time doing it as they take in weeding, gathering and piling up the weeds. The same hoe flattens the ground so it remains level. But I warn some of our farmers that tilling scorched earth excessively, is not only superfluous work, but also detrimental, for the result is that the land will be washed away by heavy rains, and therefore less productive: it is useful to have obstacles to retain the earth, and in steep terrain it is even necessary to make these holes as support from floods from higher ground.

By means of this method I get the following benefits: 1st, to keep the ground tilled; 2nd, to delay the germination of another weed, with the hole, all the roots and seeds are pulled out, which, for the most part, are useless because they are too deeply buried; 3rd, to leave the tilled soil in a state of improvement; as it is done in fruit orchards of all kinds, all the more reason to do this in tea plantations, where there is no inconvenience of the hoe cutting its roots which are deep; 4th, finally, because with this method I do not need another one. This operation, carried out in the months of June and July, prepares the land to better nourish the tea plant and produce good corn, which I have planted in a row through the gaps, placing a small portion of manure in each hole.

During these months when the tea plantation is growing, the seeds which sap the substance destined for the leaves are removed; and if it cannot be completed at this time, then they are remove at another time of the year. The defoliation operation, as taught by Friar Leandro in his pamphlet, was never adopted by me except in a small experiment, because it is contrary to the rules established by nature. The plants receive fluid through the roots, through the trunk, and above all through the leaves; to cut them before their time, is to render them useless taking the plants from their networks of nourishment; this is one of the reasons our ants dry up big trees.

I tried it on some plants. I saw that in October (in São Paulo, the month of the highest harvest), they were filled with sprouts, but I observed that when these are harvested, they produce few and paltry leaves. On the other hand, I saw that the other plants not producing many sprouts, effectively produce a harvest, which greatly favors the farmer, because if all the plants sprouted at one time, a great part of them it would be lost. I also saw that in the month of October of this year (1832), it took me a lot of work to prepare the tea because this first sprouting happened almost at the same time: it was necessary to work the ovens every day. With the exception of the overseers; there were tasks for every person to handle. In November, December and January, the harvest decreases, and it takes three days a week. From January to May, the harvest decreases further until the onset of the cold season. Nature is then idle, the plants are receiving new nutrients by means that nature provides; when the heat arrives, these substances develop into sprouts, and mature well according to the time that the cold kept them in their pods

From what has been expounded, I conclude that, for the time being, I will not use defoliation; if I am persuaded otherwise, and conclude that the gains exceed the work, then I will change my opinion. And as I am convinced of this,

I can't help but think that this notion was imparted to Friar Leandro by the Chinese tea master, who was wily enough to mislead a wise man, until he becomes aware of the deception. The ways of identifying scoundrels are not learned in books or in the school of good morals, they are only learned in the school of the ways of the world. This trickery of the Chinese matched the method of the court of Dom João VI. I know that the King wanted, and ordered, that Chinese plants be chosen and propagated; but the courtiers did not agree to this. The Inspector, despite being Brazilian, was apprehensive to upset the courtiers, fearing some sort of reprisal and subsequently his dismissal.

The facts prove this and confirm the details that the aforementioned Inspector shared with me. I saw that all the seeds were carefully collected, they were then stored until they rotted: I never even saw the tea which was made there, I know that it was manufactured behind closed doors, and I know the difficulties I had to overcome to obtain tea seeds and a cast iron furnace in China.

When the tea is not harvested, the bush is able to rise to twelve and more palms in height, which is not convenient; for it must be maintained in such a way that it does not exceed six to seven palms to facilitate harvesting; and this is achieved by harvesting it annually, so that very little growth will occur. Those who write about this plant, claim that it must be harvested after three years onwards. This is the rule; but there are some exceptions. Some plants do not develop well in three years, while others that are two years old, are ready for harvest. This is why my system recommends harvesters not touch their small plants, and that in new tea plants, only collect the buds from the twigs that already exceed four palms; then they will branch out and take on side sprouts.

All tender and soft leaves should be chosen, as well as all those that when rubbed can still be rolled up without breaking and reducing to pieces. From the long sprouts that already have four or six leaves, the top sprout with two or three leaves is removed; the others, which are tougher, are removed by cutting part of the pedicel with the fingernail, through which new shoots will sprout. This is easily learned through practice; but the farmer must know the reason for the process in order to direct people to do the job. The Chinese, who started work in this area at an early age under their parent's guidance, are very skillful in the harvesting process: they each carry their basket with three or four divisions, tucked in the left arm, deftly tossing the corresponding leaf in each division, according to the degree in which it is found to be more or less tender, since it is known that the more tender and delicate the leaf, the greater the quality of the tea. The tea garnered from the most delicate sprouts is made for the Emperor of China with great care. This is the so-called *imperial tea*, which I only saw brought to Rio de Janeiro one time by the former viceroy of India, *Cabral*.

From what has been said, it is clear that the method practiced by the Chinese master of Lagoa is not perfect: it may be that this method is used in his Province; but it's very crude, and this is not used at large in China and Japan. Besides the great inconvenience of mixing good tea with inferior tea, there is another important factor, which is: the tender leaves are cooked in five minutes and are ready for the rolling mat; the tougher leaves takes eight to ten minutes. Therefore when mixed together, the tender tea is cooked, the tougher tea is still underdone; and when it becomes ready, the tender tea will either be scorched or a little dry so it will not roll up well. This is why I never put into practice the Chinese method, and will continue to use my method to separate the leaves. I became even more committed to this notion, after sending a sample of my tea to Father Leandro, and he sent me a can of his own, I could tell by the smell, by the taste, and then visually examining the opened leaves in the teapot, that the tea was scorched, something which never happened to me, despite the fact that African boys were manning the ovens. Therefore, I conclude that the tea must go to the ovens separated by quality which avoids the detailed separation of leaves after roasting. Close to the same amount of time is spent separating the leaves before or after roasting, nothing is gained by mixing good quality tea with the inferior kind.

There are very few persons that the farmer can trust to separate the leaves. A skillful family mother will be careful to inspect the leaves cautiously. That's why I've adopted the method of bringing the leaves mixed together, and, for this task, I employ blacks, boys, women, and even children six years of age and older, who can also do some of the work. There is no general rule about the amount of leaves that may be harvested, as it depends on the output of the individual picker, on the abundance of sprouts, and, above all, on the motivation to make money. I have seen women hired for 30 Reais for each pound of leaves processed earn 420 Reais which corresponds to 14 pounds of leaves.

On the eve of the day that tea is to be made, the pickers work until the end of the day beginning immediately after lunch if the weather is cool, or at four o'clock in the afternoon, if it is hot. This tea is sprinkled with cold water and left out in the evening air in baskets or strainers. In the morning, the pickers return to the harvest, and the people chosen for separation or selection, work on the tea which was harvested the day before. For this task, the leaves are tossed onto a large table, where sprouts and tender leaves are separated for the finer and better-quality tea, and the thicker leaves for the tea that I call "Basic", which makes good *Hyson tea*. At this time, some tougher leaves and stems, and some other leaf or an insect that happens to be among the leaves is discarded. The pickers harvest until nine o'clock, when they come for lunch, and then again until one o'clock when the day's harvest is over and everyone gets involved in sorting the leaves, with the exception of those who work the ovens, which include three oven cooks, three tea mat rollers, and an individual who keeps the ovens lit at all times.

At this time the main oven cook should have the ovens cleansed, the wood ready and the area swept clean. The work begins by cooking all of the tea; which the main cook does, supplying cooked tea to the two other cooks to roll the tea with the other three, thus five persons rolling altogether. Tea, as I said, remains five to ten minutes in the oven, depending on its quality. Supervised by the cook, each one takes a handful of leaves in both hands, then proceed to the mat to rub or toss the leaves, in the manner described by Father Leandro, they squeeze the tea between their hands to extract a large amount of a greenish liquid that is so acrid, it hurts the hands of those who are not calloused. After the tea is well rolled, it is spread onto covered strainers, as described in the pamphlet by Father Leandro. The rollers than wash their hands. At that time another batch is cooked and ready, so the rollers do not remain idle. I suggest that the cook clean the oven with a thick cloth before tossing in other leaves.

I had a task put into practice for economic reasons. I ask that the excess liquid from the leaves be squeezed into a bowl, from which I get two usages: 1st, to keep the floor clean from mud created by squeezing the liquid onto the floor; and the 2nd, to capture the leaves mixed with the liquid that would otherwise be lost. After straining the liquid, these leaves are squeezed and mixed with another tea. In a large operation it is worth making the most of these residuals.

My factory has three furnaces: one of them was cast in China, and it serves as a model for casting other similar ones in the iron foundry in São João de Ipanema.

I arranged the furnaces to vent the smoke externally, leaving the oven door inside the building. Smoke bothers those inside, not only because they suffer from inhaling the smoke but also because of the heat. However, I do not approve of leaving the fire on the outside because of the bad results it could produce. The ovens are made of cast iron, and they heat up very quickly, the degree of heat of the oven is only known to those who have experience with it, so it is essential that the person in charge of the fire be close to the ovens in order to quickly increase or decrease the heat.

Dried pine branches are very good fuel for cooking the tea until it is macerated, as I explained before; from that point on, I use cordwood, the pine branches are used at times to keep the fire lit without the need to fan the flames.

Regarding the preparation of tea, I divide it into three stages: the 1st, is the cooking and rolling as explained; the 2nd, is the drying; and the 3rd, is the toasting. To dry the tea, a sufficient portion goes into the oven so that it can be slightly stirred without the danger of spilling; and this operation uses the same degree of heat with which it was cooked. The main task of the cook is to keep stirring lightly, thus preventing leaves from sticking to the bottom or onto the edges of the cooking vessel and becoming scorched. In the 2nd stage, as well as in the 3rd stage, all three ovens will be working if one oven is not enough to process all of the tea which is ready for this stage.

As soon as this herb is allowed to dry, it gives off a repulsive smell; then it exhales a pleasant aroma, similar to hay when the sun shines onto a haystack. Finally, it develops the traditional aroma of tea, at that point, the tea is considered ready; this aroma is only noticeable once the tea is dry. It is at this time that the cook orders the fire to be removed, and the tea remains under moderate heat, now the tea roasts until it acquires a gray color, the 3rd stage is now complete. I recommend that when the tea is dry and all that remains to do is to roast it, it should be taken out of the oven and given another roast, and continue until the task is finished. So, the dried tea goes back to the ovens to be roasted until it attains its color. This break is to save time, because on this occasion what has dried in two batches is roasted in one, as it is already dry and not very bulky.

After roasting, the tea is left to cool under towels, and the next day it is packaged in a box or can, as taught by Father Leandro, the important factor is that the cans should be well sealed and properly filled, if not, mold will grow over time.

After the tea has been harvested, the last operation follows, which is the final roasting, sorting and packaging. Start with the finest quality and end with the basic type. After pouring out a box or a large can, the tea is passed through a fine sieve to separate all the leaves stuck together, and sifted to find any leaf that, by chance, was not macerated and therefore did not roll up. The same is done with the inferior quality tea that has the largest portion of poorly rolled leaves. These leaves and the matted leaf fragments make another tea, which Father Leandro calls "Family Tea", which, being composed of badly rolled leaves and residuals of all kinds, does not look very good, but has a very good flavor after aging it for three years.

Since both main teas are strained through a fine sieve which only allows the residuals to pass through, another tea can still be extracted from this, I call the this tea urim; and this is done with a coarser strainer that only lets the very fine tea pass through, but not the residuals, the tea has a nice look, and it is slightly below the quality of the superior one. This part is where the last tasks and preparations is carried out by The Chinese of Lagôa, as narrated by Father Leandro in his instructive memoir.

I must warn the public that the nomenclature of teas is nothing more than a charade to take money from Europeans in the Canton market. Some are random names given by growers, other are names from the provinces of origin and from the larger factories. It was not too long ago when a kind of tea was named "Monkey tea". Since there is not a single inch of uncultivated land in China, and there are some mountains of stone, the Chinese have climbed them and planted tea through the crevices of the rocks, and because the ascent and descent of these cliffs makes it difficult for the women to harvest these crops, they have had the patience and skill to teach monkeys how to pick the sprouts and toss them downhill. This tea called monkey tea, which is sold in Canton as a rarity, is mixed with other teas harvested by hand. We all love a novelty, this is how we delude ourselves and are taken by the hype of merchants that have no faith or morality. And how many Brazilians are among us who don't like tea from Brazil, just because it is not from China! "Pearl" type tea is not always good, nor can it be, for it is made up of several types of tea, as manufactured by the Chinese of Lagôa; for example; at first sight it looks fine by the formation of the leaves, however, there are some excellent ones that are made of chosen sprouts, and that the Chinese with great dexterity roll by hand. This superfluous luxury can only be produced in China, where a worker's daily work is paid for with a quota of rice.

Tea in ball format is easy to make. A portion of good tea is taken as soon as it has been rubbed and compressed: it is pressed again by hand and formed into a ball; this ball is then placed in a piece of clean linen, and tied with a rubber band, the compression is increased. The ends of the cloth are cut, and it goes in the oven to dry with other teas, and it goes back to the oven as many times as necessary until it becomes dry and toasted. After drying the outside, the ball becomes hard, the cloth is removed and the ball is returned to the oven to dry inside. Father Leandro, who had not seen these balls, speculated that it would be necessary to add some gluten to make them. I made them without this mixture, because the tea has a lot of gluten already, and in some balls from China that they gave me in Rio de Janeiro, I could see signs of the threads of cloth with which they were tied until they dried. This kind of tea lasts a long time without spoiling: it is used in Japan where it is a luxury to serve ground tea that is mixed with a proper delicate brush. In businesses, a servant offers the tea with boiling water, another the ground tea and a brush, and the third offers crystallized sugar chunks.

I could expound further on this matter, but this is not my objective. I am happy to tell my countrymen that tea from the youngest leaves is the best tea, whatever its name or form, and that, among tea plants, the leaves of the youngest plants make better tea than those from the old plants. With this data, they can proceed any way they wish using their experience.

Having done what I said above, return the tea to the ovens in portions personally chosen: roast it sufficiently in low heat, until it is well roasted and completes its greyish color. When this operation is finished, the packaging follows; in boxes from China that are in good condition, or in large tin cans of one arroba or more, and even smaller ones if necessary. It is recommended that the tea be well packaged in a dry storage container free from strange smells, which can easily contaminate the tea. For this reason, because new tins have a scent of pitch, I usually sanitize them by filling them three times with hot water, the last time, I fill them with brewed tea. Once the last water is emptied, I have the can washed with clean water and dried with a towel, afterwards it is exposed to the sun, until it is dry, then it's brought inside. I have the inside of the can rubbed with tea residuals, and, shaken, it is then ready to be filled with tea.

It is a mistake to suppose that the tea taken out of the oven can be consumed right away, it will be strongly bitter and very unpleasant to the palate; it will not display its amber color in the water; and, above all, it is a narcotic, it is so powerful that I have seen its narcotic effect on a man who drank a lot of tea which had only aged for one year. At the end of two years it is ready, it colors the water, and produces that clarity of mind that is privy to good people. However, even at this stage of development it still possesses an herbaceous taste. That's why I only consider tea to be perfect only after three years of aging; by then it has developed its full aroma, the bitter notes are mild and the herbaceous taste has disappeared.

I don't doubt that the repetition of further roasting will shorten the maturity process, such as the Chinese do with some teas, however after making the long trek from the farmlands along the Ganges River, they will arrive spoiled at the general market of Canton. I have not experimented with this method, nor do I intend to do so.

Tea arrives in Canton from very far away, and may be more than three years old, because not everyone can promptly transport their harvest. In Canton, teas remain stored from one year to another. You see, this tea is shipped to Europe, and from there it is dispatched to different ports. Who can tell how old the Chinese tea we consume is? What can certainly be asserted is that we never consume the younger tea, most likely the old tea; and this is where the differences in goodness is established, since I truly believe, that this herb, well prepared and well-conditioned, continues to improve for up to three years; then it will reach its stationary stage, which by the rules of nature, it then begins to deteriorate.

A Frenchman who spent many years in Canton buying varieties of tea ordered by persons from many places in Europe, wrote a leaflet in which he recommended to the merchants, his countrymen, that, when buying tea in Canton, not to choose by what they see, but by how it tastes straight from the teapot, and, above all, he warns them to buy that tea that still retains the herbal taste, which proves it's still young in age, because by the time it arrives in Europe it will be perfectly aged.

What I have described is to dispel some notions that are still being championed by many persons, and above all to share with tea farmers a simpler and less complicated method of preparation. The individual experiences, communicated to the public, will bring about precision and the increase of trade in this segment, which, without a doubt, will be of the utmost importance, especially for the central provinces. I can also benefit from the experience of others!

Comments.

The tea plant thrives in all types of soil, except on land where there is more sand than soil, or where there is excessive humidity. As a general rule, all clay-like soil is good, which when tilled, is loosened with a mixture of buried vegetation. As I mentioned, tea plant roots are deep, therefore they fare well during droughts.

However, they do better in cooler areas. The shade from trees is a good, except when large roots of trees are too close to the plants and take nourishment from them.

In *Jardim da Lagôa* I only saw one type of tea; however, when transported to the mild climate of São Paulo, it has produced many variations which, in essence does not alter its preparation. It is normal that, in the wide-ranging Empire of China, also in some Provinces, because of the geographic situation, or because of other circumstances occurring in certain locations; it is not accurate to categorize tea by species, I do not believe that there are any species.

Loureiro, in his *Flora da Cochinchina*,[10] mentions three new types of tea, namely: the cochinchina tea, the Cantonese tea, and the tea that is only used for oil in the margins of Canton. Linneo mentions two species, namely: green tea and black tea. However, in general, botanists as it occurs in São Paulo and in other Provinces of this Empire, denominate the plants not as two species but as varieties of one.

Quœmpfer[11] does not refer to different species. This botanical researcher was the person who wrote most precisely about tea, and who gave Europe the best descriptions of this plant from his trips to central Japan: he claims this beverage unblocks obstructions, purifies the blood, and, above all, flushes away the calcified material that produces stones and gout. He adds that

---

[10] João de Lureiro (1710-1791), *Flora cochinchinensis* (Academia das Ciências de Lisboa, Portugal, 1790).

[11] Eugene Kaempher (1651-1716), German naturalist who was one of the first westerners to travel in Japan from 1690 to 1692. He published two works containing information from his time in Japan: a section on Japanese plants including information on the tea plant in *Amoenitatum exoticarum* (1712) and *History of Japan* (1727) published in English translation after his death.

among the tea drinkers in Japan he has not found anybody who suffers from gout or bladder stones. "I do not believe," concludes Quœmpfer, "that there is known a plant in the world, whose infusion or preparation, when taken in large quantities, is so gentle on the stomach, passes through so quickly, uplifts dejected spirits so well, and gives as much joy as tea does." Therefore, it seems that we must accept that there is only one species of this plant, and that the exotic names given to it by the Chinese is for the purposes cited earlier.

Father Leandro do Sacramento says that four pounds of leaves produces one pound of tea. This is correct; but needs further explanation. On hot days and scorching sun, the leaves are harvested lacking external moisture and even wilted. In this state, in fact, four pounds of leaves produce one pound of tea; but if the leaves are dewy, and especially if it has been raining, then five pounds of leaves are needed for one pound of tea, with small differences.

Using approximate calculations, I can tell my countrymen who intend to start a tea enterprise, that a thousand plants produce an arroba and a half of tea [48 lbs], and therefore, the farmer who has 50,000 plants, will obtain 75 arrobas [2,400 lbs] annually. This amount includes the three or four varieties that I produce, of which more than half of the total crop will always be first quality tea. A thousand plants could produce at least two arrobas [64 lbs] if we harvest them like the Chinese do. In order to increase their product, they let the sprouts grow with many leaves, and harvest them all separately. The last leaves are already so tough that they don't roll up, and from these leaves the tea that the general public drinks is made. Here you can only harvest the sprout and three more immediate leaves, the rest is no longer useful; therefore, as I stated, I harvest every day during the month of October and beyond, depending on the state of the vegetation. To be sincere, no one would buy the tea that the common Asian people drink, because our use of tea is only among the better classes, and not among the people.

There is a fact confirmed by all travelers to Asia; the Chinese and Japanese do not drink pure water. It is the general custom in all houses, rich and poor, that the first pot that goes to the fire, is a tin kettle that is filled with water, and in it is poured a sufficient amount of tea; after boiling it is kept next to the fire so it doesn't get cold, next to it there is a ceramic vase with a grip, which is used by those who want to drink their amount of tea without sugar. To mention everything about its use, it is enough to add that the native Chinese even cook their main food, which is rice, in tea water. Taking into account this general use and the immense population of China, the reader should be cognizant of the wide range use of this plant in this part of the globe, as well as the need to take advantage of the leaves that we discard.

The tea plant grows very little once annual harvesting begins after 2 years of growth. The most it reaches is eight palms [40 inches]. When not harvested, it takes on body and height. The first plant I had in São Paulo, the product of a pair of seeds which a friend brought me, secretively taking them in Jardim da Lagôa, is still active 15 years later, and measures 14 palms. It still produces some sprouts. However, I don't know what age it can attain, and it is not even necessary to know, as there is general consensus in China, confirmed by my experience, that it is best to prune the plant when it reaches 7 to 10 years. New and more plentiful burst of tender sprouts will reward properly the planter and his effort. The pruned branches remain very dry until October, and then they are good for roasting the tea. I have a large parcel that has been planted and pruned and sprouting for three years. When it produces less, it will be retired.

It is clear, that the owner of a big farm has to plant tea every year; because even when he has attained the number of plants that he must have according to his plan, from that point on he will have to renovate his old crop annually, this way he will have an annual harvest of new tea plants producing the finest quality.

Finally, I must tell my readers that the black tea I made here, and that Father Leandro mentions in his pamphlet, was the result of a lucky error in the process. Since I learned that, with low heat, I could not reduce the leaves to the point of rolling, and since I achieved this with high heat, I was convinced that, once the tea is cooked, it should be dried with moderate heat. So I did, and the result was perfect tea, but black in color. Afterwards, I used high heat until the aroma of tea became prevalent, and then the gray color appeared; however there is no difference in flavor, anyone who wants to make black tea can do so, in the manner mentioned; but in order to achieve that taste, will have to spend twice as much time drying it.

This crop is the least expensive and laborious; it does not depend on strong workers, or on so many varieties, such as sugar and coffee; boys and girls can put in a good work day harvesting and sorting, and even in rubbing and rolling, as soon as they reach puberty. Large plots of land are not required. It is most suitable for small farms close to the cities. Its process is learned in one day, watching the tasks from morning to night. My countrymen should do their calculations in any way they choose, and they will always find it profitable to cultivate this new and valuable plant, which seems to adapt itself among us better than in in its native country. It is this reality, and Brazil's geographical position, that will champion the notion that one day Europe will dispense with the need to round the Cape of Good Hope, in search of the Cantonese market.

Brazilians, let's open our eyes, and broaden our views a little when regarding our interests.

São Paulo, January 1, 1833.

José Arouche de Toledo Rendon

Economic Report on the Planting, Cultivation and Preparation of Tea By Father Leandro do Sacramento

As noted in the Introduction to this edition, tea workers and tea plants were brought from China to Brazil in 1812. In the 1820s when Father Leandro do Sacramento was made the first director of the new National Botanical Garden in Rio de Janeiro, he found several fields of neglected tea plants in the Garden. He studied the plants and located one of the Chinese workers involved in the original project. Fr. Leandro then prepared a report containing his observations of the plants and the methods for preparing the leaves taught to him by the Chinese tea master. Fr. Leandro's report from 1825 is one of the first by a western author to reveal traditional Chinese methods of tea production.

Father Leandro do Sacramento

Graduated in Philosophy from the University of Coimbra, professor of botany and agriculture, in the very loyal, noble city and Court of Rio de Janeiro, director of the gardens of the Court's Public Promenade, Botanist of the Lagoon Rodrigo de Freitas, member of the Academies Royal of Sciences of Munich, Horticultural of London, Royal Society of Agriculture and Botanic of Gand, and the Columbian Institute.

On June 4th, 1825

His Majesty the Emperor, wishing to promote everything that may contribute to the interest of the National Trade and Industry for the benefit of his loyal subjects: Orders that the State Secretary for Imperial Affairs send to the President of the Province of....., the attached copies of the Economic Report on the Planting, Cultivation and Preparation of Tea, written by Father Leandro do Sacramento; so that the same President can distribute them to the inhabitants of this Province, in the way he considers most convenient, thus facilitating the propagation of such a precious plant.

TEA

In March of 1824, when I took over the Botanic Garden of Rodrigo de Freitas Lagoon, there was in that Garden a considerable tea plantation in three upland regions, very uneven in their extent. The smallest of these three plantings was in a state of severe condition. The other two were completely abandoned, almost suffocated by wild plants, which barely allowed one to see the tea plants emerging among from them in many places. My first concern was to save the plantation, using all the resources available. I intended to publish a report about tea plant culture and the preparation of its leaves. I was convinced that the lack of knowledge on this topic was the leading cause of the poor state of the plantations of the Botanic Garden but also the lack of wider interest in growing this precious plant even though a considerable number of years had passed since it had come to Brazil. The preparation of tea leaves had already been tried. However, the process was completely unknown, and its knowledge was limited to the last remaining Chinese tea maker who had come to Brazil and perhaps a few other people. In fact, the Chinese tea maker had not been able to publish the ideas that he knew how to execute in practice, and no other person who could have written this had done so. I could not write on a subject that had been in secret until then unless I had all the notions about it, which can only be acquired by observation and repeated experience that I had not been able to achieve in the space of a few months, mainly because other tasks occupied me. I was, therefore, busy acquiring ideas that would be capable of satisfying, more extensively and precisely, the curiosity and interest of the people who would read the report that I intended to publish in the future.

In January of the current year, I received the Imperial Ordinance of January 7th of the same year, by which His Majesty, the Emperor, ordered me to prepare collections of seeds of tea, cloves, and others, to be sent to the different Provinces of the Empire, accompanied by a report written by me about their culture, manufacture, etc. From this instant on, what was in me a devotion became a sacred obligation, which I have tried to fulfil without delay, organising the present report with the ideas I have been able to acquire. In it, the farmers will have much to excuse, out of generosity on their part, for the faults they may find.

DESCRIPTION OF THE TEA PLANT

Plant of the *Euphorbiaceae* family according to Jus. Class. Polyandry Monog. of Lin.

A seminal seedling (Rich's *orthotropa*) with the radicle corresponding to the scar or navel of the seed, which develops at germination into a perpendicular root, then branching in various ways; with the plumule, or shoot, in the same direction as the root, occupying before germination a 10th part of the seed axis. It develops vertically in a single sprout and often in more than one sprout: it germinates between a month and forty days. Calyx. Perianth. Of 5 unequal tiny threads, the outer ones being smaller, concave, rounded and entire, and persisting with the fruit.

The corolla often has 5 to 6 petals and occasionally 7 or 8: the inner petals are a bit larger, more delicate and with slightly crimped and white edges; the outer petals are partially greenish on the outer side; they are all concave, with rounded tops, and when fully expanded, all petals are slightly inclined backwards and fall off before the fruit develops. It has more than a hundred stamens, attached to the base of the corolla with linear filaments, a little shorter than the petals and with the anthers arrow-shaped, exposed, of orange colour, and opening by each of its two locules longitudinally, containing pollen of minimal grains.

Pistil — Ovary oval, dressed in short fleece, exceeding the calyx, linear style of the length of the stamens, grounded in three to the upper end, and with the clefts facing backwards: 3 simple obtuse stigmas.

Pericarp — A capsule of the size of ½ inch, with three lobes, usually naked, which opens longitudinally through the back of each lobe into two valvules, and commonly contains in each cavity a globe-shaped seed attached to the axis of the capsule: the capsules by abortion, or monstrosity, vary more or less, both in the number of the lobes and of the seeds and their forms.

A perennial shrub, regularly 5 ½ feet high, garnished with many branches from the root, and therefore seems not to have a main trunk; with branches spreading variously, from which other branches up to the third and fourth order emerge, originating from the azure leaves: some of these branches frequently die, sometimes the main one, being replaced by others, usually more numerous. The leaves are alternate and hastate — elliptic, consistent, smooth, obtusely closed at the margin, with the top slightly striped and the disc marked by slightly prominent oblique veins of one inch long, with short and plump petioles on the underside. The bark is smooth, greyish and, while young, reddish at the ends of the branches.

The flowers are garnished with short, plump stalks. The flowers grow on the leaves' wings along the entire length of the last branches, in groups of 3 to 8, and, usually, a part does not survive. The plant blooms all year round, with no exception in the winter months "in the climate of Rio de Janeiro"; however, the most favourable time for the production of seeds is during the summer. It is native to China and Japan and cultivated in almost all botanical gardens in Europe, subject to particular curiosity. In Brazil, it is a plant that will soon be on the same level as coffee and sugar cane.

TEA CULTURE

The tea plant is multiplied by seeds. Perhaps it may also be multiplied by other means, as some other vegetables do, "which is very probable". However, as the multiplication by seeds is more probable and safer, and the specimens more vigorous, the multiplication by grafting, layering, etc., would be reduced to mere curiosity.

The seeds can be planted every month because in a large tea plantation, in the climate of Rio de Janeiro, there are ripe seeds in greater or lesser quantities. Moreover, the seeds (which do not retain their germination power for long) germinate well in all months of the year however new born plants will not prosper in the same way. This is due to causes that influence the vegetation, which is very varied throughout the different seasons of the year, but which can, however, be conveniently modified by industry and labour, for instance, by making shelters for the new plants when the weather is very hot, or by repeating the watering when there is a drought.

To ensure that the sowing corresponds well to the farmer's wishes, the choice of the seeds is crucial because many of them look good on the outside but have no kernel inside, either because they have not been properly fertilized or because the germ has not fully developed.

Sometimes, there are perfect seeds in the same pericarp or fruit and others that are only bark. Therefore, it is advisable to choose the perfect seeds so as not to waste time and effort on those that are abortive.

This choice is made quickly and safely by dropping the seeds into the water while they are still fresh: those that sink are good, those that float are discardable, and those that neither sink nor float completely can be used if there is a lack of seeds, in which case you need to double the number of seeds for the same area of land.

Sowing is either done in the same places where the plants are to remain or in nurseries, from where they are later transplanted. The first method can be used as an advantage when the intention is to garnish the borders of vegetable patches, roadsides, banks of streams, rivulets, etc., with these plants. In addition to having no inconveniences, it offers the following advantages: 1° these seeds, without any additional work, benefit from the care given to the surrounding plants; 2° they save transplanting work; 3° they are not damaged by transplanting. In this case, however, it will be necessary to sow every two feet, for the plants must be four feet apart from each other. One should consider those that do not grow and those that will die later, which will be gathered so that more can grow in those places. When, however, it is necessary to plant in large clumps, it will be more convenient to do the sowing in nurseries. Firstly, because it is easier to benefit from a small plot of land in a nursery than in a cluster; secondly, because the large plot of land does not remain unfruitful for one or two years, but can yield two crops of fruit for the sake of a nursery plot; and thirdly, because the plants in the nursery naturally thrive better and develop in less time.

In either case, one should choose a soil where clay or mud predominates, because some observations lead me to believe that tea does not thrive in light soil, that is, in sandy and loose soil, even if it is well manured.

In nurseries, one can sow each chosen seed 6 inches apart from each other, and at a depth of approximately one inch, in suitable soil. To prepare the ground, it should be cleared, dug with a hoe to the depth of a large span,[12] sprinkled or well turned over, and adequately levelled on the surface. Then it should be divided into beds of the desired length, with a width that allows for watering and cleaning through the side walkways.

The seeds should be put in the ground as soon as picked or within a few days because the tea seed dies when it dries out, and this happens within a few days.

---

[12] A span is a measure equal to the distance from the tip of the thumb to the tip of the little finger with the hand open in a horizontal plan, 8 to 9 inches.

For this reason, when the seeds cannot be sown immediately after being picked, it is best to keep them in slightly damp soil to retain the necessary freshness. This should also be done when seeds are transported to other places where the journey will take a few days. The seeds should be buried by hand one by one and well covered with soil.

Within 40 days or so, the seeds will have germinated if care has been taken to keep the soil properly moistened by watering if necessary.

After the plants sprout, all the care is limited to weeding and watering when necessary. Within a year the new plants will have grown to about a span, and in this condition, they can be transplanted, taking care not to damage their roots, because although they are not very delicate to transplant, the energy of their growth depends on this.

The 6-inch distance at which the seeds were planted in the nurseries is sufficient to be able to transplant the new plants with their respective root ball, so as not to damage the roots. The best time for transplanting is in July and August, because it is not too hot and also because the circumstances regarding the state of the soil are more favourable. At this time, the soil is neither too dry nor too wet, and at these extremes, the plant hardly grows, even with its root ball, which means more work.

After transplanting to the places where the plants will remain, which must happen at a distance of 4 palms from each other, it is necessary to water immediately even if the soil is humid, because through watering it is possible to adjust the soil to cover the roots, preventing them from drying out to the detriment of the plant.

When you choose to practice the second method, which is to sow directly in the places where the plants will remain, and taking into account what was mentioned regarding planting in nurseries, it will be sufficient to bury the previously selected seeds at a distance of 2 palms from each other. By this method, those at a distance of 4 palms should remain while plants in between will give enough to transplant in case of failures.

If it is necessary to use doubtful seeds, that is, those that did not go to the bottom of the water and did not stay on the surface, it is definitely necessary to throw in each hole 3 or 4 seeds instead of 1, to increase the probability of success of the sowing, and in places where more plants grow, to pull out the weakest and leave the strongest.

Subsequent care in the cultivation of tea consists of weeding, which should be carried out as many times a year as the state of conservation and cleanliness of the land requires, because by doing it regularly you can count on a good plantation.

All the grass that is removed is gathered in small piles between the plantations and, after it has withered, holes of 3 palms deep are dug in the gaps between the plants, burying the grass in them, which should be covered with earth to a depth of a palm. This practice allows the cleaning of the plantation and the improvement of the soil: it is necessary, however, to take care not to bury either plants as Fragant Flatsedge (*Scirpus odoratus* of Linn.) or Indian Shot (*Canna indica* of Linn.), as well as other plants known as Fragrant Flatsedge or Ferns, because these plants grow buried at greater depths.

Watering should be done when planting requires it. Irrigation can be practiced through furrows, if there is enough water for it, or through sprinkling with watering cans or equivalent, to be adapted to the dryness of the weather.

In the second year and sometimes at the end of the first year some of the new plants flourish and give seeds. If these are needed to enlarge the plantation or for new plantings, it is recommended to keep them. Otherwise, all the flowers should be removed as soon as the small buds appear, so that the plant becomes more vigorous and does not have to share its nutrition with the flowers and fruits, which absorb the largest and best quantity of the plant's juices.

When the plants reach a growth of 3 palms and are garnished with plenty and vigorous branches, which happens in the third year (and may happen earlier in some climates, where they prosper better) they are in conditions to be harvested, compensating all the care and work given to them.

It is, however, necessary, or at least advantageous, to prepare the plant before the leaf is harvested.

This preparation consists of removing all the old leaves from the plants, so that only the green, healthy leaves remain.[13] This defoliation can be done well or badly: to be well done, all the leaves should be removed one by one, leaving the stem of the leaf, and even some of the leaf attached to the stem, without damaging the tender sprouts that may exist. This operation can be done by all kinds of invalids, except blind people with poor tactile skills: it is advisable to leave the leaf stem on the plant, as it will encourage the new sprout, which will soon develop on its base or upper part. At this time, you should remove all the existing flowers and fruit, so that the plant uses all the juices that it would be divided

---

[13] In his report on tea production, Arouche questions this practice of tea plant defoliation. He suggest that the Chinese tea master recommended the practice to damage the plants and thereby reduce competition with Chinese tea producers. Modern practice is to keep leaf on the plant to support production of new shoots for plucking. This and other modern management practices allow plucking about every two weeks during the growing season. -*ed*

between the flowers and fruits to germinate new sprouts, which should be harvested for the production of tea.

Plants prepared in this way will in 15 days produce leaves for harvesting. These leaves are those that already existed in embryo and by staying on the plants accelerated their development through the abundance of juices, from which they have exclusively benefited. The bad defoliation method consists of closing the hand with which each branch is held at the bottom and sliding the closed hand up the branch, tearing off all the leaves, flowers and fruit at once. This operation is very quick but inefficient, because by doing this you also remove the existing sprouts, as you can easily see. In this case, only after 30 days, a little more or less, will the plants have new leaves for harvesting, and even these in smaller quantity, as a result of the destruction of the sprouts that were beginning to germinate and the irregularity with which the old leaves were torn off, because some were torn off with the foot and others were left with larger portions of flaps than would be convenient.

It is advisable to defoliate the plants days before spring, according to the variations in different climates. In this province of Rio de Janeiro, the right time for this task is October or the end of September. After this defoliation once a year, the plants continue to give new leaves during the whole period of warm temperature, which in Rio de Janeiro lasts for 6 to 7 months, from

October to April.

If plants are defoliated later, harvests will be smaller and proportional to the extent of warm weather that follows. In winter, the plants rest and strengthen themselves for the following year. According to Kempfer, it is said that in Japan the bushes are cut off near the root every 7 years. This seems very appropriate to me, to give origin to new and powerful branches that result in abundant and better-quality harvests. But as this has not yet been done here, I am mentioning it for when it is practised.

After the first task of preparing the land, all the other tasks can be carried out by weak people, a great advantage of tea planting.

The application of manures has a great influence on the results of this planting; they should of course contribute much to the growth and vigour of the plants, but I fear that they may cause bad results in the quality of the tea, especially if they are animal manures. In this respect I have no experience yet, but as soon as it is done it will be possible to decide with certainty.

TEA HARVESTING

When the plant has sprouted new sprouts, and these have already four to six leaves, these new leaves should be harvested, as they are more tender than the old leaves, which are hard, brittle and dark green.

Women, boys, and even imbecile people are able to harvest this crop, which is done by hand, picking the whole leaves one by one and cutting with the fingernails the tender sprouts, where two to three of the most tender leaves are found. When cutting the tender sprouts, one should be careful to leave a few, which as they grow form new stems. These should be adequate in number and location to the vigour of the plant, as the tea plant is a shrub where some stalks often dry out and are replaced by others. A simple glance at the plant is enough to see when and where it is best to let new sticks grow, so that the shrub will remain vigorous and regular in form.

The leaves that are being harvested are placed in a basket hanging on the left arm of the person. The harvest starts at dawn and ends at midday, during which time each person harvests at least two pounds of leaves. It is not difficult to harvest three pounds or more, depending on the skill that one has acquired through practice.

Harvesting is done in two periods of time: the first one, from the beginning of the day until 9 o'clock, lunch time; the second one, after lunch, until midday. The leaves collected in the first period by the different pickers are gathered together and immediately spread out in a large and uncovered flat basket inside the house or in the tea workshops, so that they do not get too hot from being piled up too high or exposed to the heat of the sun. The same applies to the leaves picked in the second period, and it is convenient that the place where these leaves are deposited is small and ventilated.

The harvested leaf must undergo three preparations to reach the state of perfect tea, which I will now explain.

FIRST PREPARATION OR PRODUCTION OF TEA

Taking a portion of about 2½ to 3 pounds, starting with the leaves that were picked before lunch, they are thrown into the cooking vessel or *wok* [14] of the oven, the temperature of which must be the same as that of the cassava flour ovens. A worker, sitting on the edge of the oven, successively stirs with both hands the leaves inside the said vessel until they have sweated enough and become wilted. This operation will take about 10 minutes, being complete when the leaf yields easily to the wringing without breaking, which is tried on 3 or 4 leaves of the wok. Then remove them immediately with a vessel or basket without leaving any leaves in the wok, for which a small broom should be used. The remains should be thrown into the same vessel, placed under the

---

[14] In the original document the Portuguese word '*caldeira*' is used to indicate the vessel for cooking tea. This translate as a 'boiler' however the vessel described here corresponds to what modern readers would recognize as a wok. See figures 6 and 12 in Bruce's *An Account of the Manufacture of the Black Tea* (1838).- *ed.*

outside edge of the pan and held with the left hand, while the right hand removes with the broom the rest of the leaves that have remained in the wok.

Once the leaves have been removed from the wok, they are immediately thrown onto the matting in two piles of similar size, placing another equal proportion of leaves in the wok. These should undergo the same first withering operation. While this second portion of leaves is being scalded in the wok, two men take each his portion of leaves which had already been scalded and thrown on the matting, on a table, and holding them with both hands, still hot, rub and crush them with force against the matting, successively, for 10 minutes, until all the leaves have been equally exposed to a degree of twisting which mechanically alters their tissue to the extent that a considerable part of the leaves are crumbled. When the leaf is thus prepared, it will release its juices to the point of making the workers' hands wet. The leaves are then taken in this state, in small portions, to the point of pressing them comfortably between the palms of both hands, and rubbed lightly between the palms in a circular motion, dropping in each movement a small portion of the leaves. In this operation, the leaves that have already been crushed into a cake form divide and spread evenly, and it is possible to see if all the leaves are crushed. If not, then they should be brought together again and the rolling should continue until all the leaves are crushed, concluding the operation by rubbing them between the hands as explained above.

The operation of crushing the leaves against the mat is similar to that of kneading flour dough to make bread, or that of some washerwomen, who, after pouring soap on the clothes, rub them with both hands against the stone, or board, where they wash the clothes.

When this operation of rolling up the scalded leaves is concluded, as explained, the leaves are spread out or scattered in a basket to clear off the matting. Now the second portion of leaves that had been scalded in the wok may be prepared or rolled up. This operation is repeated until all the leaves have been scalded and rolled.

The first leaves, which have already been scalded and kneaded on the mat and placed in baskets, are completely cooled down while you do the same with other portions of leaves that have lost some of their moisture. Then take 2 or 3 portions from the first baskets, and throw them all at once into the wok, stirring in the same way, without interruption. Every care should be taken to ensure that the leaves remaining at the bottom of the wok are always replaced by others, for two reasons: 1st, so that the leaves that are hotter come to the surface, or even mix with those that were on the surface (and therefore less hot), and distribute the heat equally between them all, in order to promote evaporation uniformly; 2nd, to prevent the leaves remaining at the bottom of the wok from burning or tasting burnt, which would cause the whole batch to be lost. This operation continues for approximately 20 minutes.

When this portion of leaves is drained, but not yet dry, a new portion of crumpled leaves can be thrown in, especially if there is a large quantity of leaves to prepare, because when these leaves are again mixed with those already dry in the wok, they dry out more quickly. It is easy to see that by mixing them with a larger portion of already dried and heated leaves, they quickly acquire the same degree of heat, which promotes the evaporation of their moisture. Or one can remove the dry leaf from the wok and spread it in another basket, where it will continue to evaporate part of the moisture it still has, throwing into the wok a new portion of the crumpled leaf, which should be given the same treatment already mentioned. And so on until all the leaves taken hot from the wok are given the degree of first drying, which are then spread out in baskets, where they continue drying.

After drying all the crushed leaves in the wok, as described above, the previously dried leaves that had been cooling in baskets are thrown back into the wok, and twice as many leaves are thrown into the wok, continuing to stir in the same way until the leaves are dry and toasted. One should apply this operation, which will take from ½ to ¾ of an hour, on all the leaves harvested that day, taking care to reduce the fire of the furnace as the leaves dry out or become toasted to avoid the danger of being burned if the fire is not lowered or if you stop stirring.

While still in the first degree of roasting, the nice smell of the tea starts to be noticed, becoming more intense and pleasant as the degree of roasting increases. Through it, you can judge the degree of roasting at which this operation should end, and this point is reached when the herbaceous smell is barely noticeable. The leaf prepared in this way is already tea you can use, and even if you still notice a grassy aftertaste, it is less unpleasant than the taste of old tea.

After this first preparation, the herbaceous flavour of the tea can be significantly reduced by prolonging the roasting in the wok, for example, for another hour or 1½ hours. Still, it is not advisable to do so. In the first place, there is a risk of the leaf acquiring the smell of roasting. Secondly, because it is difficult to extinguish the whole herbaceous flavour in this first preparation, to the point of being able to say that the tea is good, since it is a mixture of leaves of different qualities, more or less tender, more or less rolled, some whole and others almost powdered. Naturally, the leaves that are best rolled, the smallest and most powdered leaves, because they always occupy the bottom of the wok, are exposed to a higher degree of roasting than they need to be. At the same time, the older leaves, which roll up more easily and naturally occupy the upper part of the wok, even when well stirred, will not be subject to the desired degree of roasting because they only touch the bottom of the wok slightly, which results in the herbaceous flavour. This flavour is always transferred to the other leaves, even if prepared at the right roasting degree. It is impossible to stir equally a mixture of fragments unequal in nature, gravity or weight, so it is always convenient to use other means. For this purpose, either the leaves that were subject to the preparation of the first day are kept and added to the other portions prepared on the following days until a considerable quantity of about one arroba[15] is obtained. Then the subsequent operations are carried out, or these operations are carried out the next day on the portion prepared the day before, even if it is small. It is also possible to do this on the same day, taking care to sift the leaf that has been reduced to powder. However, what is advantageous is to add the tea prepared on the first day to the preparations of another day or days when the arroba or arrobas of tea in this first state has already been reached. This convenience will become evident when the following article describes the other preparation process after which the tea is partly ready.

After being exposed to the preparation of the second day, tea can be preserved for a certain number of years,

---

[15] An arroba is an older measure of weight equal to 32 pounds in Portuguese usage

becoming better as it ages. In this first state of preparation, it is strongly consumed by the middle classes of China: the subsequent preparations to separate the different qualities of tea which occur in the trade are normally carried out in other workshops, which deal positively with this objective. The small Farmers compete to sell small portions of tea only with the first preparation. Its value is judged according to the second examination made by the buyers, who check the quantities and different qualities, being already so trained in this that they can hardly be mistaken. For this reason, they rarely disagree on the prices.

In this first state, the buyers of what could be called "raw tea" are those who, from tiny portions of different values, prepare large quantities of the different qualities of the tea available in the market.

These historical ideas are based on the knowledge of the Chinese tea master of Jardim da Lagoa, who told me that before he left his homeland, he was involved in this market from which he made his living. I found what I heard very credible, as it was in accordance with the facts that I will present next.

In 33 pounds of tea — raw tea — that I prepared, brewed to the state where it reaches perfection, I only managed to separate ½ pound of pearl tea, all the rest were teas of different qualities. The large boxes of pearl tea which come from China ordinarily contain 90 pounds. One would therefore need 180 arrobas of raw tea to obtain a box of pearl tea. It cannot, therefore, be a product for a farmer, still less a mediocre one, if one considers the customs and way of life of the Chinese, according to the ideas we have of them. It seems that large bales of tea of certain qualities, such as pearl or *aljôfar* (which is only a variety of pearl) can only be obtained from large quantities of tea, probably prepared in workshops where many portions of tea compete, as the Chinese say. This is especially true if we consider that the tea picked from the plant, after the first preparation, loses almost ¾ of the weight it had as a fresh leaf, and that the harvest only takes place a short time during the year. In addition, no matter how skilled a man may be, he will not be able to pick 8 pounds of fresh leaf in one day, which is reduced to about 2 pounds of tea of different qualities, the pearl or aljôfar being contained in only one ounce.

SECOND PREPARATION OF TEA

The second preparation of the tea begins with the division or separation of the different qualities and conditions of the leaves, which, being all mixed, as said in the first preparation, constituted what I call raw tea. This separation is done as follows:

A portion of raw tea of about three pounds is taken and thrown into a sieve made of bamboo or 'taquaruçu', an equivalent material, whose screen is of medium size. The raw tea is separated into two portions by conveniently shaking the sieve between both hands. These portions consist of what passes through the sieve, shaken over a sizeable shallow basket of tightly woven cloth. The other remains on top of the sieve, each of these portions being about half of all the raw tea. This procedure is continued with equal quantities, separating all the raw tea equally, putting together in one basket all that remains above the sieve and in another all that passes through it. The leaves that pass through the sieve are all those that have been reduced to small fragments by the operation of rolling and stirring in the wok, and also the smallest and most tender leaves that have been well rolled. Those that remain on top of the sieve are the larger leaves, and together with the leaves that are older than the appropriate age for rolling up, they will not be rolled up.

Once this first separation of the raw tea into two portions has been made, about two pounds of the thick tea that remained on the sieve is taken and thrown into a shallow basket of well-covered cloth. It is similar to that practiced when one wants to separate the husk from piled rice or coffee by cleaning it after pounding, which is called — quibandar (to sift). And by this way, one separates the leaf that has not rolled and its fragments, which will not pass through the sieve because they are lighter. If some pieces remain, this operation is continued on all the leaves that were left in the sieve.

This operation is done on a large shallow basket of tightly woven cloth, where all the leaves that were not rolled and all the tiny fragments or leaf dust fall. Only the rolled-up leaf remains in the sieve, already separated and cleaned of all the dust or small pieces that escaped from the sieve.

After separating the leaves that remained on top of the sieve, all the sieved leaves are thrown into the wok conveniently heated, in portions of about 3 pounds at a time, and successively stirred in the wok for more than ¾ of an hour. This operation should be practised for all leaves of this quality.

In this state, the leaf does not run the risk of burning, as it is separated from the dust that would always occupy the bottom of the wok. In this condition, it is subject to a more homogeneous roasting. After passing all the leaves to be treated through the wok twice and cooling them in baskets, they are sieved again on another coarser sieve. Everything that passes through this second sieve is deposited in another shallow basket of woven cloth, and *ying-chen* tea being separated through a fine sieve from the powder formed by stirring and roasting in the kettle, is ready after this second operation to be drunk.

By separating the leaves through the second sieve, two qualities are obtained in two different baskets: the *Ying-chen* tea and the thicker leaves that remain on top of the second sieve. All well curled and rounded leaves are manually selected, which constitute the pearl and aljôfar tea. The remaining leaves give origin to the tea called 'Chutho-Thé', of the same quality as pearl tea, differing only from it for its less curled leaves. It is ready to store.

At first sight, picking the tea by hand, leaf by leaf, seems to correspond to a task so boring and time-consuming that anyone who is not meticulous, as the Chinese are supposed to be, would be discouraged. But this operation, in reality, is not so meticulous as it seems, for it is done on a number of leaves where all the thin leaf and dust has already been separated. The latter constitutes one tenth of the total amount of the raw tea, with about one sixth of the leaves being used to make Chutho-Thé tea. This operation is quite comparable to the practice of the bakers of vetch or chaff from wheat, who hand-pick large portions of wheat, without causing discouragement to the bakers. A woman in less than 1½ hours chooses all the pearl or aljôfar tea contained in an arroba of raw tea.

The selected pearl tea returns once more to the wok in a moderate degree of heat, where it is stirred until it is smooth. Then it is sieved through a thin screen through which the powder formed in this last operation is separated.

This pearl tea, thus prepared and ready, can be further divided into aljôfar tea by passing it through a sieve with a screen capable of separating the coarser grains from the thin ones. The latter ones make the aljôfar tea.

The tea separated from the first time sieved raw tea, i.e., that which has passed through the first sieve, contains two qualities of tea. These two qualities are separated through the fine sieve through which only the dust passes. The fine rolled tea that remains on top of the sieve is returned twice more to the wok over low heat until it has acquired a greyish colour. After this, it is separated from the dust that is again formed by the roasting through the fine sieve, and this is the so-called Hysson tea, which is then ready. This is about one third or the raw tea, as is also the case with Ying-chen tea.

If you want to separate the Uchim tea, take all the powdered tea separated from the different qualities of tea, which has been gathered in the same basket, together with the non-rolled leaves already sieved, and separate the finest powder with a very fine sieve. When the first sieve has also divided the non-rolled leaves, the middle leaf between these two extremes is called Uchim tea, which does not need to go any further in the wok, and which, being separated from the suitably roasted leaf, is completely ready. The separation of Uchim tea must not be a common practice (as is usually the case in China) for the following reason:

When I explained how to separate the different qualities of leaves from the raw tea, I specified that once all the leaves had been separated into two portions with a medium sieve, the non-rolled up old leaves were separated by sifting and added to the small fragments that escaped the sieve. If one separates the Uchim tea, only the unrolled leaves remain, mixed with a small portion of other leaves of different qualities of tea which should not be added to the Uchim tea. Therefore, these leaves, or waste, would be of little or no use on their own if mixed with the Uchim tea. As the finest powder would be useless if separated from the Uchim. These leaves are combined to form a quality of tea called Thó tea. Although it is of inferior quality, it is produced in large quantities and commonly used since it makes up about a quarter of the raw tea. I have heard that it is the most widely used tea in China, so it may be advantageous in Brazil as it is more economical. This tea could be called a 'family tea'.

In distant countries where all the tea-Thó is not consumed and where a large quantity of tea is exported for trade, it may be advisable to separate the Uchim tea.

The different qualities of ready tea are each kept in their respective jars, which should be made of lead or tin and covered well. When more portions are prepared, and you want to pack them, you must give them a degree of roasting to preserve the tea better.

THIRD AND FINAL OPERATION

This last roasting is the third and final operation and is done as follows. Put all the Chuthó-thé tea in convenient portions in the sieve, and sort the remaining unrolled leaves and any dust that may have been formed in the roasts of the previous operation. This Chuthó-thé tea is roasted again at moderate heat for between ½ and ¾ of an hour. After this, it is sieved with a thin screen to separate the dust again formed by the roasting. And so, it is ready and, in a state, to be packed.

In the same way, the Ying-chen tea is passed through the sifting to completely sort out any remaining fragments of non-rolled leaves and the dust formed during the roasting of the preceding operation. Once the tea is completely clean, the powder from the Hysson tea is also removed with the thinnest screen. Once the Ying-chen and Hysson teas are clean, mix them and roast over moderate heat for ½ to ¾ hours, stirring carefully to not crush the leaves. After roasting, separate one from the other again with a sieve of a suitable screen for this separation. Once separated, pack each of these qualities of tea. The same is done with Pearl tea, which should then be sieved through a thin screen to separate any dust that may have formed.

All the unrolled or poorly rolled leaves and dust or fragments, which have been sorted in this third operation from all qualities of tea, constitute a quality called Chuthó tea, which is superior to Thó tea and inferior to all other qualities. From this quality of tea, the finest dust and the largest unrolled leaves are sorted, and the Uchim tea can be formed, which is of better quality than the Nahim tea when separated from the Thó tea. This quality can be added to this Uchim tea separated from the Thó tea, which will result in a medium quality Nahim tea.

China does not make this separation for the same reason that it does not make it for the Thó tea. And both the Thó and Chu-Thó teas are good and no weaker compared to the Hysson teas that come from China.

There are no other operations to be carried out apart from packing each quality of tea in its lead or tin container. It is advisable that these containers can be filled and well covered so that the tea is well preserved.

REFLECTIONS ON TEA CULTURE

The land on which the tea plant is grown in the Botanical Gardens of Rodrigo de Freitas Lagoon is a sunny and dry but fresh plain situated about two fathoms above sea level. The nature of its soil is a mixture of clay, which is not homogenous. In those places where the first shrubs were planted, the soil is slightly better due to repeated digging and some manure, which has been applied there. However, the predominance of clay means that the soil, when dry, is quite hard, as is typical of clay soils.

The tea bushes in these first places are the most vigorous: their average height is 5 feet, and their age is almost 14 years.

Some of these shrubs, sheltered by some trees which protect them without shading them, are over seven feet high and very vigorous. However, in other places where less care has been taken in digging, weeding and manuring, the ground is much more compact and the shrubs much less developed.

In three successive years, I have sown tea seed in the Garden of the Public Walk of this Court and have obtained lush plants that wither in summer despite the watering and manuring conveniently applied. In this ground, sand predominates, and the night dew is not frequent, especially in the summertime, when the tea plants have perished, for the most part. The same fate has befallen other plants grown in other places where the soil is light, i.e., sand predominates, although there is sufficient manure.

It seems essential for the prosperity of tea growing that the soil is clayey. Its compactness completely shelters the plant's roots, providing them with humidity, which is characteristic of clayey soils. The affinity between clay and water is such that even the heat of metal smelting cannot deprive the clay of all the water it contains. In fact, clay soil is the one that retains and preserves humidity the longest, despite its dry appearance on the surface when there is a lack of rain. On this basis, the clay pyrometer is constructed. With this, the degree of heat of smelters and glassworks is assessed.

Among the leading causes of the Province of Rio de Janeiro's fertility, where one sees an eternal verdure even when there is a lack of rain, is the predominance of clay, which constitutes almost all its territory. It comes from the decomposition of feldspar in its mountains of granite and chalk. And the fact that Rio de Janeiro suffers the lack of limestone transformations for mortar constructions, which is solved by the calcareous products of the sea worms, is overcompensated by the incomparable advantage of its soil fertility.

I do not intend to attribute fertility exclusively to clay soils and sterility to calcareous and sandy soils. This would be a gross error since even calcined limestone is a good mineral manure in many cases. However, given the circumstances of the geographical locality in which Brazil finds itself, if the predominance of its soil were calcareous and not clayey, it would be sterile, in general terms. It would be fertile only with regard to an incomparably smaller number of plants and localities.

I am, therefore, convinced that clay or mud soil is particularly suited to the tea plant. However, the proportions and state of the combination of the clay with other substances in the soil, which we may call clayey, are prodigiously variable. Moreover, it is not only the chemical nature of the soil that influences the vegetation but also many other causes of equivalent energy, such as temperature and its variations, exposure, the degree of humidity, the pressure of the atmosphere, and meteoric action, between others. Thus, the different tea plantations should vegetate with varying degrees of energy according to the advantageous circumstances in which they find themselves.

Although I have no experience of the energy of the tea plant vegetation on the slopes of hills and in their gorges, it seems that I can say without fear that the tea plant thrives better near woods or forests where the evening dew provides a daily watering, even if small, in glens, in the gorges of hills, on the banks of rivers, streams and brooks not subject to diurnal flooding or other similar conditions than in circumstances different from these.

The tea plant is a shrub that maintains perennial vegetation, modified in different degrees of energy in the different seasons. It is always green and blooms even in winter (in the climate of Rio de Janeiro). It is suitable for a warm and humid climate, so almost the whole territory of Brazil is favourable for the cultivation of this plant, and perhaps it can even be cultivated in the Province of Rio Grande do Sul. Only by experimenting will we know.

Tea growing is more appropriate to the climate of Rio de Janeiro than to that of China if cultivated in those places where it is grown as the first and most important branch of agriculture. The reason for my thinking is that they grow four crops a year in China, while in Rio de Janeiro, they grow six. This is the same as saying that the time of the most significant energy of the tea plant vegetation in China lasts four months, while it lasts six months in Brazil. An advantage of 50 to 100, in favour of Brazil over China.

There are in China some varieties of the same tea plant that are not currently found in Brazil, one of them being the variety of plants from which the so-called white teas are made. Its leaves are garnished with fleece on the underside. However, let us consider that these varieties are unknown in a province and characteristic in other remote places. It seems safe to conclude that these varieties are due to the influence of different climates. Consequently, as this plant is cultivated in Brazil, we can expect numerous variations. The present memoir was already at the printer's when I received a letter from the Province of S. Paulo, written by the Honourable Marshal Arouche. I transcribe the following paragraph because it contains news relating to tea, which confirms my assertion that the tea plant, naturally, must suffer variations due to the different climates of Brazil. The paragraph is as follows: — Here, I certainly have three qualities reproduced from a new seed. The most common plant has a medium-sized, lance-shaped leaf. There is another tea with larger and more rounded leaves and a slightly darker green, and there is a third tea whose bush grows less and has such a small leaf that it looks like common myrtle. These variations should naturally multiply according to the climatic variations, which I think will be confirmed by experience.

To support this theory, I would like to mention the following fact. Marshal Arouche wrote to me from São Paulo telling me that he had obtained black tea when he tried the first trials of tea brewing. He blamed this on the cooking pan of the iron furnace, which is certainly not true, as it is also the case with the wok at Lagoa, or on an error in the process, about which he asked me for exact knowledge, etc. This information leads me to presume that the difference in the result obtained by Marshal Arouche is due to the climate variation and the plantation condition because, according to his testimony, the shrubs in São Paulo were bigger and more vigorous than in the Botanical Garden of Lagoa, despite being much younger. However, I do not dare to affirm this as a certainty because it is impossible to narrate the whole process in detail in a simple letter and only by knowing it would I dare to confirm my opinion.

But whatever the case, the example of the Marshal is, fortunately, being followed by others, who have already begun to plant tea, as he tells me in the same letter. I hope that soon tea will be as common in Brazil as coffee and tobacco, but this will not happen while this culture is limited to samples. Therefore, our farmers must undertake the cultivation of tea in the same way that they cultivate manioc, sugar cane, coffee, etc. As the useful product of this plant consists of the tender leaves and the leaves are small, extensive plantations are necessary to pick considerable quantities of leaves daily and obtain profitable harvests.

All vegetables whose useful part is the leaves are more profitable than those cultivated for their fruits, and therefore their harvests are less contingent. The veracity of this position is so evident that I think it is not necessary to prove to him that the cultivation of tobacco, indigo, and all the oleaceous plants, etc., will never cease to be advantageous to those devoted to them, provided nothing subversive affects those cultivations.

The tea plant occupies one of the most distinguished places among the plants cultivated for the interest of its leaves. For this reason, in the countries where it thrives, the extent of its cultivation is greater than that of any other vegetable. Indeed, 40 million pounds of tea are exported daily from China, and the quantity consumed there is incalculable. How many people are needed for this crop? Although the wages of tea workers are very modest, the fact that so many people are employed in this crop is undoubted proof that the profit it generates is greater than that of many other occupations.

This advantage of China and Japan cannot fail to be applied to Brazil, where tea's prosperity is not inferior to those in those empires. Thus, tea cultivation in Brazil promises a secure livelihood for the people who engage in it, even if they are millions. It means a real advantage for the nation, whose strength depends on the number of individuals who have a secure livelihood.

REFLECTIONS ON THE TEA BREWING PROCESS

The tea preparation essentially consists of: 1st - dry scalding of the leaves; 2nd - mechanical alteration of the organic tissue of the leaves; 3rd - boiling and dry roasting at a suitable degree of heat; 4th - separation of the different qualities of leaves prepared, which constitute the different qualities of tea.

The dry scalding of leaves — the last stage of improvement for which the most advantageous instruments have been invented — is done in an iron wok with a mouth three feet in diameter and seven inches deep, oval in shape, similar to the pans in which chandlers melt the wax for making candles. In fact, the shape and capacity of this vessel are the most suitable and advantageous for this purpose, as a man or woman, sitting on the edge of the furnace where the wok is placed, turn with his or her hands all the leaves submitted at a time to the scalding, without sudden but regular movements. The oval shape of the wok is particularly suitable for this operation because as the leaves all fall to the bottom it is enough to move the bottom of the wok to turn them without having to touch the bottom of the wok, which would cause burns to the worker's hands. As the leaves do not stick to the wok, light pressure on the upper part is enough to move them to the sides, and others take their place. The scalding is completely done when the leaves, having sweated enough, yield easily without breaking to the twisting or curling when tested on two or three leaves. This operation takes about six minutes, depending on the degree of heat and the number of leaves. The degree of heat is approximately the same as when cassava flour is cooked. If it gets so strong that it starts to burn the worker's hands, one must reduce it to the degree the worker can bear without discomfort.

Given this, it is easy to understand that the scalding operation can be performed on any other vessel, with more or less advantage to the operation and the operator, and in more or less time. A clay pot made in the model format mentioned above will do very well, provided it has a very smooth internal surface. However, it will be necessary to apply more heat to the furnace for clay heats less than iron, as is the case with clay ovens for baking flour compared with copper. A copper or iron pot could also be used, but as the pot bottom is flat, this does not favour stirring. In this case, it will be necessary to speed up the operator's movements to ensure that the leaves do not burn due to lack of stirring.

Scalding by fire is quite different from that practised in the sun, and even if the leaves were reached the same degree of withering in the sun, the alteration would be very different and, as such, fruitless, as is easy to see.

The mechanical alteration of the organic tissue of the leaves is the second part of the tea brewing operation. In this operation, the aim is to crush all or a large part of the vessels that form the leaves tissue so that they roll up in the furnace's heat and the juices contained on it, when they overflow and mix, reach the state necessary for the chemical alteration. This operation results in the formation of new products and the disappearance of existing ones. In order to better conceive the course of nature in this, I will now recall facts comparable to this one, and which are quite trivial.

Some bunches of grapes or oranges, for example, when preserved sound and perfect, are free from alterations other than withering and will eventually pass away or dry up if they do not rot. These fruits are crushed until most of their skins rupture, and the juices spill out and mix. The result will soon be fermentation and the formation of alcohol, which would never happen if the fruits were not mechanically altered. Therefore, if one intends to make wine from grapes, oranges or other juicy fruits, one begins by mechanically altering their tissue to expose their juices to alcoholic fermentation. In the same way, it is necessary to alter the tissue of the tea leaves when one wants to obtain leaves ready for that degree of alteration that is called tea. Grapes are crushed with the hands or feet. The tea leaves are crushed with the hands but could also be crushed with the feet if their quantity requires it, as is reportedly the practice in some provinces of China. The essential thing is the crushing. The most useful tool for this is the matting described on the previous pages. This matting, however, is not so essential that it cannot be replaced by a stool or table, although with less advantage.

By this crushing, one does not obtain the leaves curling but is predisposed to it, for the curling is a spontaneous and necessary result of the roasting heat under such circumstances.

The third essential (and perhaps most important) part is the boiling and roasting. This operation consists of throwing the already crushed and still damp leaves into the hot wok at the same degree of heat they were scalded, stirring continuously, as when scalding.

The degree of humidity that the leaves must be in is not indifferent, which means that if the leaves are left to sweat for a long time after being kneaded, the baking result is not so good for two reasons. Firstly, because the drier leaves cannot withstand, without deteriorating, the degree of heat of the first cooking, which is more potent but well withstood by the leaf due to its humidity. This is necessary for the leaves' juices to change and give rise to other juices that give the tea its pleasant aroma and reduce the bitter flavour. The dissolution state of the leaves' fixed principles in the water of their vegetation still existent contributes significantly to this. Secondly, during the time needed for the already crushed leaves to reach a greater degree of dryness, no further spontaneous alterations of the same juices could be suspended, for nature does not rest for a single instant in her operations. Therefore, the juices would already be slightly different, as would, consequently, the result of the cooking. This theory seems confirmed by facts, as it is said that the tea which in some parts of China is brewed in the sun after being scalded in the fire and bruised, is essentially different from the tea brewed and roasted in the fire, one difference consisting in the lack of smell.

As soon as the tea smell is perceived, it is necessary to reduce the heat degree, either because the desired change has already been obtained or because the leaf, already deprived of moisture, no longer needs so much heat. Also, there could be a danger of burning the leaves and causing discomfort to the worker.

Tea prepared using the first-day process retains an unfavourable herbaceous flavour, and the more pronounced this flavour is, the better the quality of the tea. This flavour diminishes progressively with repeated slow roasting, to such an extent that the first quality tea is the one which is roasted the most often and for the longest time in the oven. This flavour could be entirely extinguished by the repetition or duration of the roasting process. The China tea master, after preparing the first portion and considering it as ready, in the first qualities of tea, one could still perceive a remnant of an herbaceous smell, which had disappeared entirely in the tiny qualities of Chá-thó and Chu-thó. The China tea master himself assured me that this is characteristic of new teas, and I did not want to interfere since it was the first quantity of tea being prepared. Being Director of the Garden, I did not want to risk anything because I was anxious to have the honour of presenting His Lordship, the Honourable Emperor, samples of prepared tea. However, I have the following comments to make on this:

Tea-Thó and Chu-Thó teas are made from the oldest, less rolled leaves and the small fragments or dust of the first quality teas. Since this dust is of the same nature as the first quality teas, why does it completely lose its herbaceous smell?

Because during roasting, they have occupied the bottom of the wok longer. So, if you do the same thing for a longer period to fine teas, they will naturally present the same results. There is another reason why I have not yet attempted the progress of roasting in fine teas, which is that the Chinese purposely use this way of roasting in fine teas for very valid reasons. Among them is that teas in this state can be preserved for a long time, progressively improving as happens with good wines, which, while new, are less drinkable than those of inferior quality. When ready for use, the inferior teas may not keep for a long time. Only through experience can we be sure of this.

The roasting on three different days does not influence the result of the preparation and is done only for convenience reasons. I am convinced that the preparation process, which started on one day, could continue without interruption until the different qualities of the teas are separated.

The instruments used in boiling and roasting are baskets, where the leaves taken out of the oven are spread out to cool while others are roasted.

Any clean vessels that do not spill or leak leaves through their fabric or texture could be used for this, such as towels laid out on tables.

The separation of the different qualities of leaves after roasting is an operation that is partly essential and partly accidental. To better understand this it is necessary to reflect that the crushed leaves are unequal in age and size, though generally they are young leaves. In what is improperly called rolling, some remain more whole and others more broken when crushed. In this operation, some leaves are reduced to tiny fragments, and all these together go to dry and toast in the wok, where the leaves capable of doing so are rolled up due to the oven's heat and the stirring. Many leaves do not curl up, and almost all the minor fragments remain the same. As soon as these fragments reach a sufficient degree of dryness, they occupy the bottom of the wok. By the continuance of roasting, they come to roast themselves entirely because their surfaces are larger than their solidity and the space they naturally occupy. The same happens with the leaves that do not curl up because they are bigger. Suppose they are entirely roasted on the first day of operation and mixed with the rolled leaves that still need a higher roasting degree. In that case, they must be separated so as not to suffer excessive roasting that would result in their loss and the loss of the other rolled leaves before they reach the necessary roasting degree. Therefore, this separation is indispensable so that the leaves that are not rolled and those which are separated and reduced to dust give the quality of Cha-thó tea. For the same reason, before the operation on the third day, the leaves rolled and roasted on the second day should be sifted to separate the dust that is again formed by stirring. This dust, together with the leaves that are also separated through sifting, form another quality of tea called Chu-thó, and it is essential to sift the dust whenever the roasting process has to be repeated on different days.

Sorting the different qualities of rolled leaves after they are ready is a random but essential thing because of the custom of consumers to buy teas with homogeneous leaves.

I assume that some qualities of tea prepared in China, in which one perceives scents different from the scent of tea, are artificial products comparable to the well-known method to create different scents when different qualities of snuff from different factories are added together. The most appropriate teas for these confections should be those that, not having been roasted over an open fire, are deprived of the aroma of green teas and are more capable of acquiring the aroma that one wishes to attribute to them. In this case, as the scents are compatible with the tea, and in moderate amounts, they seem natural, as happens with snuff. The tea known as *Chulon*, in which a certain smell of roses is clearly perceptible, seems to have been prepared with such an artifice.

This theory of mine could arouse the curiosity of many people who will want to try experiments on this process, which I have not had the opportunity to try.

The varieties of tea in bale, bundles or piles, of which I have only heard, must be produced by appropriate manipulations. For example, if one chooses the most tender leaves and, instead of rolling them, crushes them in a mortar and reduces the paste of these crushed leaves to pellets or bales, subjecting them to dehydration and roasting between tea leaves, one may perhaps obtain the tea in bales. This tea should be of superior quality because it includes tender leaves, and its texture is equivalent to the maximum degree of curling. The leaves that are thus crushed can remain compacted during roasting by perhaps adding some substance that gives them a certain degree of firmness without altering the smell or taste. For this purpose, any pure mucilage, such as gum Arabic, could be used in a convenient quantity. I have never seen these types of tea, and therefore I am unaware of their context and everything else that I would have to examine in order to be able to advance my conjectures on this matter.

TOOLS THAT MAKE UP A COMPLETE TEA PREPARATION WORKSHOP

- A house with a capacity for six men to work in it with one or two ovens. These ovens have no ashtray distinct from the laboratory and are built of tile, clay, or lime, with on one side the door through which the fire is lit, and in the middle, an iron cooking vessel well joined to the furnace, so the smoke does not escape. The shape of the furnaces is that of a parallelepiped three feet high and four feet long and wide. The cooking vessel, oval in shape, is three spans wide and seven inches deep. A wooden rim attached to the vessel's outer side is needed to ensure that the leaves do not fall out when roasted. This rim is similar to the rim around a sieve, of the height of a cross-hand and can be made of American muskwood (*Guarea Trichilioides* of Lin), which is very common and very suitable for this and similar works. These cooking vessels are sold in China and cost a Spanish peso, according to the information of the Chinese tea master.

— A basket of cloth similar to the Mozambique mats cloth, made of bamboo, or *taquaruçu*, or *taboca* (a species of Bamboo from Lin). The basket edges is garnished with a wooden rim of American muskwood and the height of a cross-hand, with a flat bottom and five spans in diameter.

This basket is used to sift the leaf when dividing the tea qualities, and to hold harvested leaf, which must be spread out to keep it fresh.

— Two more baskets, similar to the one above, with the only difference that one is four spans wide. The tea is spread out in these baskets when it is taken out of the oven to cool: they are also used to sift the tea when it is divided into qualities, and also to spread out the freshly picked leaves.

— Two baskets, made of the same fabric as the preceding ones, with a diameter of three spans in diameter with a cross-hand woven bottom: these baskets are used to receive the qualities of tea already separated so that each quality does not mix.

— Two baskets of a shape similar to an egg broken vertically into two equal parts and truncated at its thin end, or in other words, of a semi-cucurbitaceous shape, truncated at the end. These baskets are two palms long and six inches long at the truncated end. Their fabric is regular for baskets but very tight and made of bamboo or *taquara*. The margin is thick because a reinforcement of curved wood garnishes it. The truncated top is not reinforced. They are used to measure the amount of leaf to throw into the wok and mainly to remove leaf from the wok when necessary.

— A sieve (a kind of screen which serves to clean through ventilation, so its fabric is closed or covered) garnished with a wooden rim one inch high.

— Four sieves with square screens. The thickest one does not allow Pearl tea to pass through. The other screens are progressively smaller, the last one only letting fine dust through.

— A small vessel 6 inches long and proportionately thick to sweep up the tea that remains in the wok.

— A mat 6 palms long and 3 palms wide, garnished at its edges with a reinforcement done of timbouva tree (which may be of another material) for consistency, and a handle at each end, which hold it on a table, fastening it under the table with a stick that is passed through the handles. The fabric of this mat is comparable in everything to that of a mat from Angola. However, it is made of bamboo or *taquaruçu*, and its fabric is woven in longitudinal ribbons. This mat is used for crushing or rolling up the tea leaves.

— Two tables: one as big as the mat and the other bigger.

— A number of deep baskets proportional to the number of people who pick the tea leaf.

Afterword

Antoine Guillemin's life was cut short in 1842 just short of his 46th birthday and a mere three years after his expedition to Brazil. The following posthumous tribute to Guillemin's life and work written by a co-worker and friend Antoine Lasèque was published in the Annales des Science Naturelles *where Dr. Guillemin had served as an editor.*

Notice on the Life and Works of Antoine Guillemin, D.M.P.

Naturalist Assistant at the Paris Museum of Natural History; member of the Philomatic Society, the Physics and Natural History Societies of Geneva, and the Curieux de la Nature in Bonn, etc.

A new loss must now be added to those that science has had to mourn in recent days. On January 15, 1842, Guillemin died in Montpellier. He was still young, but had succeeded in devoting his life to the study of natural sciences, particularly botany. They say that it is only after losing those we love that we truly come to appreciate their merits. But Guillemin's friends didn't need the influence of death's sad prerogative to recognize the true worth of their friend!

Antoine Guillemin (not Jean-Baptiste-Antoine, as he erroneously called himself) was born at Pouilly-sur-Saône, in the canton of Seurre and arrondissement of Beaune (Côte-d'Or), on January 20, 1796.

He studied at the municipal college in Seurre, where he was considered one of the most distinguished pupils. Upon leaving school he was placed with a lawyer. He worked there for eighteen months, but his interest in chemistry and desire to obtain a commission as a military pharmacist, at a time when it was difficult to avoid conscription, caused him to abandon the study of law. In 1812, he was apprenticed to a pharmacist in Dijon. After two years in that city, he went to Geneva. It was in 1815 that he first met Mr. De Candolle. Guillemin was already inclined towards the natural sciences, and lessons with this great master decided his vocation. His enduring passion for the study of plants dates from this time. It was a passion that nearly proved fatal; one day, while collecting plants in the Alps, he fell and broke his right arm. The fracture was a serious one: the injury was slow to heal, and for a time it was feared that amputation would be unavoidable. Guillemin was fortunate enough to escape the threat of that cruel operation, but the accident left him with a permanent stiffness in the elbow joint.

In 1820, Guillemin decided to go to Paris and settle there permanently. He wasn't well-known in the city, but he presented himself under the patronage of Mr. De Candolle. It was then that Mr. Benjamin Delessert offered him a position working alongside Mr. Achille Richard, who had been entrusted with Delessert's botanical library and herbariums, collections which have since been greatly expanded. Guillemin saw that this was his destiny. Indeed, where else would he have found such an excellent opportunity? The post offered everything he needed to complete his botanical education and perfect his knowledge of the very branch of natural history to which he most wanted to dedicate himself. He was also encouraged by Mr. Benjamin Delessert's warm welcome. He received other marks of favor as time went on. Guillemin sincerely appreciated this, and he often took pleasure in recalling his early days in Paris; those happy memories were filled with gratitude and veneration for Mr. Delessert! The sentiments were deeply and keenly felt, and one could say that they remained a part of him until his dying day.

He was also faithfully devoted to Mr. De Candolle. The lessons he learned from that famous Geneva botanist never faded from his mind. Guillemin never spoke Mr. De Candolle's name without also expressing his sincere devotion to the learned professor to whom, in a way, he owed his entrance into the world of science.

Mr. De Candolle was in a better position than anyone to judge Guillemin's natural abilities. Early on, he had foreseen and declared that the young student would become a distinguished botanist. Guillemin showed real affection for his teacher, an affection that the professor returned in equal measure. Far from forgetting that bond, it would seem that Mr. De Candolle counted on its lasting long after he himself was gone. As his death approached, he remembered Guillemin, and, since according to the natural order of things his student ought to outlive him by many years, he bequeathed to him, in his will, the exclusive right to publish a new edition of his *Théorie élémentaire de la botanique* [Elementary Theory of Botany]. When it was first published, this remarkable book opened a new path for science, and it will long remain a monument to the genius and philosophical mind of its author. Guillemin accepted the task with heartfelt gratitude: it was a final precious legacy of the learned man whose name will go down in the annals of science. He felt that he must reproduce the work in its entirety, and would have regarded it as a desecration to alter the text by making even the slightest change. Perhaps he would have ventured to add a few notes at the end of the volume, or at least included an explanation of a few more recently-introduced terms in the language of botany. He had also expressed his intention to add an account of the life and works of Mr. De Candolle. Who better than his pupil to perform such a task! Who could recount the fullness of his life and the great value of his work better than Guillemin! Alas, he ran out of time.

Behind the most modest of exteriors and the simplest manner, Guillemin hid a highly cultivated and extensively informed mind. He was always prepared to share the treasures of erudition within, in the kindest and most self-effacing manner. He was consulted by both the ignorant and the learned. Countless ingenious insights and elevated ideas were gathered from his conversation! Many people came to draw from this well of knowledge. The carefree nature that pervaded his manners masked the prodigious memory that was his special gift. He was able to call to mind any work or plant he had ever seen, even many years later, and he was unfailingly able to refer to these objects, which he had seemingly glanced at without paying the slightest attention, whenever needed. In plant identification, he was guided by the accuracy and rapidity of his own practiced eye. Anyone who had the pleasure of visiting Mr. Benjamin Delessert's vast botanical collections could not help but be delighted with their interactions with Guillemin, the assistance he provided, and the extreme kindness he showed to each and every visitor. In 1827, he was appointed Botanical

Assistant at the Paris Museum of Natural History, a position he held while continuing to work for Mr. Benjamin Delessert. He brought to his new job all the same qualities that distinguished him in the other. The two roles helped him form a natural connection that was very beneficial to science between the national museum in the Jardin des Plantes and the private museum of Mr. Delessert.

Guillemin contributed several essays on plant organography and physiology to the scientific world, as well as works on descriptive botany. These include, among others:

Considérations sur l'hybridité des plantes en général [Considerations on the Ability of Plants in General to Hybridize], published conjointly with Mr. Dumas. In this essay, the authors discuss the phenomenon of hybridization, which they had observed in certain alpine species of gentians. While they give no decided opinion on the theory of hybridization, they do indicate the peculiar circumstances under which wild plants cross-breed.

Recherches microscopiques sur le Pollen [Microscopic Research on Pollen]. This essay was published prior to, and may have been the inspiration for, the research that learned physiologists have been conducting on pollen in recent years. Guillemin explains its general structure and anatomical composition with clarity and precision. He points out its similarity of form in various genera of a certain number of natural families, a similarity that could be used in characterizing the families.

Flore de Sénégambie [Flora of Senegal-Gambia], published with Mr. Achille Richard and Mr. Perrottet.

Mémoire sur le Pilostyles [Essay on Pilostyles], a curious new genus of that most singular family, Rafflesiaceae.

Zephiritis Taitensis, a short work that contains interesting information on the botanical geography of the Society Islands, principally the island of Tahiti, and their vegetation and flora.

In addition to these various scientific works, a complete list of which can be found at the end of this notice, his collaboration in numerous collections and dictionaries devoted to natural history is attested to by a wealth of articles, scattered throughout various publications. For five years he was the main editor of the botanical section of Férussac's *Bulletin universel*. He taught botany for three years at the Institut horticole in Fromont, including his renowned *class on Botany and Vegetable Physiology*. In 1833, he founded the *Archives de botanique* [Archives of Botany], a well-executed compilation published with the generous assistance of Mr. Benjamin Delessert. It was later merged with the 2nd series of the *Annales des Sciences naturelles* [Annals of the

Natural Sciences], which Guillemin edited jointly with Mr. Adolphe Brongniart from then on.

Guillemin laid down his pen, never to pick it up again, after describing in the *Annales* the characteristics of *Jaubertia*, a new genus of plant. At Mr. Benjamin Delessert's suggestion, he named it after a former minister and member of the Chamber of Deputies, Count Jaubert. Even the most serious duties could not distract the Count from botany, and he found solace from the rough world of politics in the peaceful study of plants.

Guillemin had a pronounced gift for working and writing, and he could have left us a greater number of works, but his essentially distracted nature did not allow him to focus for long periods of time on a single supervised project. The sustained concentration demanded by serious endeavors was strangely exhausting to him. He may have refused to publicize his scientific understanding, but he shared it with everyone around him. If he hadn't, what a terrible waste it would have been of such a mind, filled to overflowing with a reliable and diverse array of knowledge!

In the openness of intimate conversation, Guillemin showed his true nature in a very particular light. He could be playfully mocking, and a somewhat sarcastic philosopher, but there was no malice in him. He would cheerfully let slip, in the middle of a conversation, a variety of anecdotes and lively incidents that his memory held in reserve, gathered who knows where. He was full of benevolence for all, and was never envious. No fame ever overshadowed him. His criticism was without bitterness. He seized upon and exposed the ridiculous wherever he found it, not violently, but with clever, good-natured playfulness.

In July 1838 Guillemin was appointed by the Ministry of Commerce and Agriculture to go to Brazil to study the cultivation and preparation of tea, and to bring back plants that the government hoped to naturalize in France.

He left Paris on August 10, 1838 to travel to Brazil, immediately after being accepted as a pharmacist by the School of Pharmacy in Paris. When he arrived in Rio de Janeiro, he carried out his mission as carefully as possible. One year later he brought back to France 18 crates, containing the 1,500 surviving tea starts that remained out of the 3,000 plants he had acquired in Brazil and the ripe seeds he had had sown in the spaces between them. He also brought a great number of samples of precious woods for dyeing and cabinet-making, and a host of substances used in making medications, the precise applications of which would be valuable to the trade. As a result of this mission, which he recorded in detail for the Minister in a report submitted upon his return, he received the decoration of the Legion of Honor. The reward honored the intelligence and zeal he had displayed on his mission as much as it did his knowledge and previous work, which had already earned him a place among the most skilled botanists.

Lately, Guillemin's health had begun visibly deteriorating. He was suffering from an organic disease, and his friends' advice was of no avail in convincing him to attempt to stop its swift advance. Although he was named a medical doctor in 1832, he had never been particularly interested in the study of disease. That may explain his skepticism of medical treatment. However, the alarming state of his health finally compelled him to consult a physician. It seemed that warmer climes might help his organs regain the energy they had lost, so on January 2, when it was a mere 6 degrees outside, he left his friends behind. He went to Montpellier seeking the milder temperatures that Paris couldn't offer at that time of year. Unfortunately, when he arrived in Montpellier he discovered that, contrary to all expectations, the weather there was even more intensely cold. The journey he had undertaken was very difficult in such a season, but he had endured it without too much fatigue. In Montpellier he was delivered into the officious care of highly experienced physicians as well as that of the most affectionate of friends. There were so many reasons to hope! Delighted to learn of his nomination as a member of the School of Pharmacy in Paris, he was full of confidence in the treatment he had just undergone. "One more month," he wrote, "and my recovery will be assured." Only a few days later, death carried him away from his family, his friends, and science!

Mr. Benjamin Delessert always felt a deep interest in Guillemin, and the announcement of the tragic news affected him terribly.

All those who knew Guillemin will long remember him. I was his fellow worker and colleague at the botanical museum of Mr. Benjamin Delessert. He called me his friend, and for ten years, I had the good fortune to spend every day with him. In writing this notice, I hope to pay a final tribute to his memory, and to bear witness, as far as I can, both to my sincere regrets and to the deep emptiness that his loss has left in me.

List of Works of A. Guillemin, Related to Organography and Plant Physiology, and Descriptive and Applied Botany

1. Essai d'analyse chimique de la racine de *Gentiana lutea*. Report co-authored with M. Jacquemin, d'Arles. (*Journal de Pharmacie*, avril 1819).
2. Considérations sur l'hybridité des plantes en général et particulièrement sur celle de quelques Gentianes alpines. Report co-authored with M. Dumas, de

l'Institut. (*Mémoires de la Société d'histoire naturelle de Paris*, 1823, vol. 1, page 79, avec une planche coloriée).

3. Notice sur une monstruosité des fleurs de l'*Euphorbia Esula*, (*Mémoires de la Société d'histoire naturelle*, vol. 1, page 93).

4. Mémoire sur l'organisation du fruit des Cistinées, et particulièrement sur celui des *Helianthemum*. (Lu à la Société Philomatique, le 20 mai 1825). Unpublished.

5. Recherches microscopiques sur le Pollen et considérations sur la génération des plantes. Lues à l'Institut, le 21 mars 1825. (*Mémoires de la Société d'histoire naturelle*, 1825, vol. II, page 101, with a colored plate),

6. Observation d'une monstruosité de fleurs du Lilas vulgaire. (*Mémoires de la Société d'histoire naturelle*, 1828, vol. IV, page 363).

7. Note sur les affinités du *Joliffia africana*. (*Mémoires de la Société d'histoire naturelle*, 1827, vol. III, page 320).

8. *Icones lithographicæ plantarum Australasiæ rariorum*. In-4°. Paris, 1827. 20 planches.

9. *Floræ Senegambiæ tentamen*. Co-authored with MM. George Samuel Perrottet and Achille Richard. Tome 1er; in-4°. Paris, 1830-1833. 72 plates.

10. Plantes grasses, peintes par Redouté. Le texte des 29e et 30e livraisons, publié en 1832.

11. Description du *Dombeya Ameliæ*, nouvelle plante d'ornement. (*Annales de Fromont*, tome III, mars 1832, with an engraved plate).
12. Considérations sur l'irrégularité de la corolle des *Calceolaria*. (*Archives de botanique*, tome II, page 1).
13. Mémoire sur le *Pilostyles*, genre nouveau de la famille des Rafflésiacées. (*Annales des Sciences naturelles*, 2° série, vol. II, page 29, with an engraved plate).
14. Mémoire sur les effets de l'enlèvement d'un anneau circulaire d'écorce sur la tige du *Pinus sylvestris*. Lu à l'Institut, en février 1835.
15. Observations sur les organes microscopiques, communément appelés vais seaux poreux dans le tissu du bois des Conifères. Lues à l'Institut, le 12 décembre 1836. (Supplement to the previous report).
16. Considérations sur l'amertume des végétaux, suivies de l'examen des familles naturelles où cette qualité physique est dominante. Thèse soutenue à la faculté de médecine de Paris, le 30 août 1832 ; in-4°. 59 pages.
17. Coup-d'œil sur la végétation des cinq grandes parties du globe. Inséré dans le *Traité de Géographie* de M. Adrien Balbi. Paris, 1833.
18. Notice nécrologique sur M. Henri Cassini. (*Bulletin de la Société Philomatique*, décembre 1832. — *Archives de Botanique*, tome 1er, page 462, with a facsimile of his writing).

19. Dictionnaire classique d'histoire naturelle ; 16 vol. in-8°. Paris, 1822-1830. — Most of the articles on flowering botany and some general articles, such as Botanical Geography, Degeneration of Organs, Hybridization, etc.
20. Bulletin universel des Sciences sous la direction de M. de Férussac ; 1821 à 1831. — The main editorial staff of the botanical section.
21. Dictionnaire des drogues simples et composées, ou Dictionnaire d'histoire naturelle médicale, etc. — Co-authored with MM. Alphonse Chevalier and Achille Richard. 5 vol. in-8. Paris, 1827-1829. — Guillemin wrote almost all the botanical and natural medical history articles.
22. Annales de l'Institut horticole de Fromont. 4 vol. de 1829 à 1833. — Plusieurs notices surdes plantes nouvelles du Jardin de Fromont et son *Cours de Botanique et de Physiologie végétale.*
23. Archives de botanique. 2 vol. in-8. Paris, 1833. — Founder and main editorial staff.
24. Zephyritis Taitensis. Énumération des plantes découvertes par les voyageurs dans les Iles de la Société, principalement dans celle de Taiti. (*Anal. des Sciences nat.*, 2° série, vol. v1, p. 297.)
25. Tableau synoptique des plantes les plus usitées dans l'économie et la médecine domestiques du Brésil, par M. le docteur L. Riedel. Extrait, revu et annoté, d'un

ouvrage publié en portugais à Rio-de-Janeiro. (*Annal. des Sciences nat.*, 2° série, vol. XII, p. 212).

26. Icones selectæ plantarum, editæ a Benj. Delessert. In-4°, vol. 111. Paris, 1837. — Diverses observations botaniques ; descriptions de deux genres nouveaux : *Martiniera* et *Byrsanthus*, et de quatre espèces nouvelles d'Eriocaulon.
27. Rapport à M. le ministre de l'agriculture et du commerce sur sa mission au Brésil. 1839. (Revue agricole, 16ᵉ livraison).
28. Observations sur les genres *Euryale* et *Victoria*. (*Ann. des Sc. nat.*, 2° série, vol. XIII, p. 50).
29. Note sur la plante qui a servi de type au genre *Bobua* DC. et sur les affinités de ce genre. (*Ann.*, id., vol. XV, p. 158).
30. Description du *Jaubertia*, nouveau genre de la famille des Rubiacées. (*Ann.*, id., vol. XVI, p. 60).

Guillemin's family felt that his wishes would be best fulfilled by donating his herbarium to the natural history museum in Dijon. A few rare species were removed and given to the Paris Museum of Natural History, according to Guillemin's specific instructions.

www.ingramcontent.com/pod-product-compliance
Lightning Source LLC
Chambersburg PA
CBHW050320120526
44592CB00014B/1987